Wi-Fi Integration to the 4G Mobile Network

Wi-Fi Integration to the 4G Mobile Network

André Perez

iSTE

WILEY

First published 2018 in Great Britain and the United States by ISTE Ltd and John Wiley & Sons, Inc.

ISTE Ltd
27-37 St George's Road
London SW19 4EU
UK

www.iste.co.uk

John Wiley & Sons, Inc.
111 River Street
Hoboken, NJ 07030
USA

www.wiley.com

Library of Congress Control Number: 2018931217

British Library Cataloguing-in-Publication Data
A CIP record for this book is available from the British Library
ISBN 978-1-78630-173-4

MIX
Paper from responsible sources
FSC® C013604

Contents

List of Abbreviations

3GPP	*3rd Generation Partnership Project*

A

AAA	*Authentication Authorization Accounting*
AAA	*Authenticate and Authorize Answer*
AAD	*Additional Authentication Data*
AAR	*Authenticate and Authorize Request*
AC	*Access Category*
ACK	*Acknowledgment*
AES	*Advanced Encryption Standard*
AF	*Application Function*
AGC	*Automatic Control Gain*
AH	*Authentication Header*
AID	*Association Identifier*
AIFS	*Arbitration Inter-Frame Space*
AKA	*Authentication and Key Agreement*
AM	*Acknowledgement Mode*
A-MPDU	*Aggregate MAC Protocol Data Unit*
A-MSDU	*Aggregate MAC Service Data Unit*
ANDI	*Access Network Discovery Information*
ANDSF	*Access Network Discovery and Selection Function*

ANQP	*Access Network Query Protocol*
AP	*Access Point*
APN	*Access Point Name*
ARP	*Address Resolution Protocol*
ASA	*Abort-Session-Answer*
ASR	*Abort-Session-Request*
AUTN	*Authentication Network*

B

BCC	*Binary Convolutional Coding*
BCE	*Binding Cache Entry*
BID	*Binding Identifier*
BPSK	*Binary Phase-Shift Keying*
BSS	*Basic Service Set*
BSSID	*BSS Identifier*

C

CCA	*Credit-Control-Answer*
CCA	*Clear Channel Assessment*
CCK	*Complementary Code Keying*
CCMP	*Counter-mode/CBC-MAC-Protocol*
CCR	*Credit-Control-Request*
CE	*Congestion Experienced*
CHAP	*Challenge Handshake Authentication Protocol*
CK	*Cipher Key*
CN	*Correspondent Node*
CNA	*Correspondent Node Address*
CoA	*Care-of Address*
CoT	*Care-of Test*
CoTI	*Care-of Test Init*
CRC	*Cyclic Redundancy Check*

CSD	*Cyclic Shift Diversity*
CSMA/CA	*Carrier Sense Multiple Access/Collision Avoidance*
CTS	*Clear To Send*
CW	*Contention Window*
CWR	*Congestion Window Reduced*

D

DA	*Destination Address*
DAD	*Duplicate Address Detection*
DCF	*Distributed Coordination Function*
DEA	*Diameter-EAP-Answer*
DER	*Diameter-EAP-Request*
DF	*Don't Fragment*
DFS	*Dynamic Frequency Selection*
DHCP	*Dynamic Host Configuration Protocol*
DIFS	*DCF Inter-Frame Space*
DNS	*Domain Name System*
DOI	*Domain of Interpretation*
DRB	*Data Radio Bearer*
DSCP	*DiffServ Code Point*
DSMIPv6	*Dual-Stack Mobile IP version 6*
DSS	*Data Sequence Signal*
DSSS	*Direct Sequence Spread Spectrum*

E

EAP	*Extensible Authentication Protocol*
EAPOL	*EAP Over LAN*
ECE	*ECN-Echo*
ECN	*Explicit Congestion Notification*
ECT	*ECN-Capable Transport*
EDCA	*Enhanced Distributed Channel Access*

EHSP	*Equivalent Home Service Providers*
EIFS	*Extended Inter-Frame Space*
EMSK	*Extended Master Session Key*
eNB	*evolved Node B station*
EPC	*Evolved Packet Core*
ePDG	*evolved Packet Data Gateway*
EPS	*Evolved Packet System*
E-RAB	*EPS Radio Access Bearer*
ERP	*Extended Rate Physical*
ESP	*Encapsulating Security Payload*
ESS	*Extended Service Set*
E-UTRAN	*Evolved Universal Terrestrial Radio Access Network*

F

FA	*Foreign Agent*
FAA	*Foreign Agent Address*
FBE	*Frame-Based Equipment*
FCS	*Frame Check Sequence*
FID	*Flow Identifier*
FQDN	*Fully Qualified Domain Name*

G

GAS	*Generic Advertisement Service*
GEK	*Group Encryption Key*
GI	*Guard Interval*
GIK	*Group Integrity Key*
GPRS	*General Packet Radio Service*
GRE	*Generic Routing Encapsulation*
GTP-C	*GPRS Tunnel Protocol Control*
GTP-U	*GPRS Tunnel Protocol User*

H

HA	*Home Agent*
HESSID	*Homogeneous Extended Service Set Identifier*
HNP	*Home Network Prefix*
HoA	*Home Address*
HoT	*Home Test*
HoTI	*Home Test Init*
HR	*High Rate*
HS2.0	*Hotspot 2.0*
HSS	*Home Subscriber Server*
HT	*High Throughput*

I

IARP	*Inter-APN Routing Policy*
ICMP	*Internet Control Message Protocol*
ICV	*Integrity Check Value*
IDFT	*Inverse Discrete Fourier Transform*
IE	*Information Element*
IEEE	*Institute of Electrical and Electronics Engineers*
IETF	*Internet Engineering Task Force*
IFOM	*IP Flow Mobility*
IK	*Integrity Key*
IKEv2	*Internet Key Exchange version 2*
IMSI	*International Mobile Subscriber Identity*
IP	*Internet Protocol*
IPSec	*IP Security*
ISAKMP	*Internet Security Association and Key Management Protocol*
ISM	*Industrial, Scientific and Medical*
ISMP	*Inter-System Mobility Policy*
ISRP	*Inter-System Routing Policy*
IV	*Initialization Vector*

K, L

KCK	*Key Confirmation Key*
KEK	*Key Encryption Key*
LAA	*Licensed Assisted Access*
LAN	*Local Area Network*
LBE	*Load-Based Equipment*
LBT	*Listen Before Talk*
LCID	*Logical Channel Identifier*
LDPC	*Low-Density Parity Check*
LLC	*Logical Link Control*
LMA	*Local Mobility Anchor*
LMAA	*LMA Address*
LMD	*Local Mobility Domain*
LTE	*Long-Term Evolution*
LTF	*Long Training Field*
LWA	*LTE-Wi-Fi Aggregation*
LWAAP	*LWA Adaptation Protocol*
LWIP	*LTE/WLAN radio level integration with IPsec tunnel*
LWIPEP	*LWIP Encapsulation Protocol*

M

MAA	*Multimedia-Authentication-Answer*
MAC	*Medium Access Control*
MAC	*Message Authentication Code*
MAG	*Mobile Access Gateway*
MAPCON	*Multiple-Access PDN Connectivity*
MAR	*Multimedia-Authentication-Request*
MCC	*Mobile Country Code*
MIC	*Message Integrity Code*
MIMO	*Multiple Input Multiple Output*
MIP	*Mobile IP*
MME	*Mobility Management Entity*

MN	*Mobile Node*
MNC	*Mobile Network Code*
MO	*Management Object*
MPTCP	*Multi-Path Transmission Control Protocol*
MSDU	*MAC Service Data Unit*
MSISDN	*Mobile Subscriber ISDN Number*
MSK	*Master Session Key*
MSS	*Maximum Segment Size*
MU	*Multi User*

N, O

NAI	*Network Access Identifier*
NAS	*Non-Access Stratum*
NAT	*Network Address Translation*
ND	*Neighbor Discovery*
NSWO	*Non-Seamless WLAN Offload*
OCS	*Online Charging System*
OFCS	*Offline Charging System*
OFDM	*Orthogonal Frequency-Division Multiplexing*
OPI	*Offload Preference Indication*
OSA	*Open System Authentication*

P

PAD	*Peer Authorization Database*
PBA	*Proxy Binding Acknowledgement*
PBCC	*Packet Binary Convolutional Code*
PBU	*Proxy Binding Update*
PCC	*Policy and Charging Control*
PCO	*Phased Coexistence Operation*
PCRF	*Policy Charging and Rules Function*
PDCP	*Packet Data Convergence Protocol*

PDN	*Packet Data Network*
PGW	*PDN Gateway*
PLCP	*Physical Layer Convergence Protocol*
PMD	*Physical Medium Dependent*
PMIPv6	*Proxy Mobile IP version 6*
PMK	*Pairwise Master Key*
PN	*Packet Number*
PPA	*Push-Profile-Answer*
PPDU	*PLCP Protocol Data Unit*
PPR	*Push-Profile-Request*
PS	*Packet-Switched*
PS	*Power Save*
PSDU	*PLCP Service Data Unit*
PSPL	*Preferred Service Provider List*
PTK	*Pairwise Transient Key*

Q, R

QAM	*Quadrature Amplitude Modulation*
QoS	*Quality of Service*
QPSK	*Quadrature Phase-Shift Keying*
RA	*Receiver Address*
RA	*Router Advertisement*
RAA	*Re-Auth-Answer*
RADIUS	*Remote Authentication Dial-In User Service*
RAR	*Re-Auth-Request*
RC4	*Rivest Cipher*
RD	*Reverse Direction*
RFC	*Request For Comments*
RIFS	*Reduced Inter-Frame Space*
RLC	*Radio Link Control*
ROHC	*Robust Header Compression*
RRC	*Radio Resource Control*

RSN *Robust Security Network*
RSRP *Reference Signal Received Power*
RSSI *Received Signal Strength Indication*
RTA *Registration-Termination-Answer*
RTO *Retransmission Time Out*
RTR *Registration-Termination-Request*
RTS *Request To Send*
RTT *Round Trip Time*

S

SA *Source Address*
SA *Security Association*
SAA *Server-Assignment-Answer*
SACK *Selective Acknowledgment*
SAD *Security Association Database*
SAR *Server-Assignment-Request*
SeGW *Security Gateway*
SGW *Serving Gateway*
SIFS *Short Inter-Frame Space*
SKA *Shared Key Authentication*
SPD *Security Policy Database*
SPI *Security Parameter Index*
SPR *Subscription Profile Repository*
SSID *Service Set Identifier*
ST *Slot Time*
STA *Session Termination Answer*
STBC *Space-Time Block Coding*
STF *Short Training Field*
STR *Session Termination Request*
SU *Single User*

T

TA	*Transmitter Address*
TAI	*Tracking Area Identity*
TC	*Traffic Class*
TCP	*Transmission Control Protocol*
TEID	*Tunnel Endpoint Identifier*
TFT	*Traffic Flow Template*
TID	*Traffic Identifier*
TIM	*Traffic Indication Map*
TK	*Temporary Key*
TKIP	*Temporal Key Integrity Protocol*
TLS	*Transport Layer Security*
TLV	*Type, Length, Value*
TMK	*Temporary MIC Key*
TPC	*Transmit Power Control*
TSC	*TKIP Sequence Counter*
TTAK	*TKIP-mixed Transmit Address and Key*
TTL	*Time To Live*
TTLS	*Tunneled Transport Layer Security*
TWAG	*Trusted WLAN Access Gateway*
TWAN	*Trusted WLAN Access Network*
TWAP	*Trusted WLAN AAA Proxy*
TXOP	*Transmission Opportunity*

U

UDP	*User Datagram Protocol*
UE	*User Equipment*
UICC	*Universal Integrated Circuit Card*
U-NII	*Unlicensed-National Information Infrastructure*
UP	*User Priority*
USIM	*Universal Services Identity Module*

V, W, X

VHT	*Very High Throughput*
VoLTE	*Voice over LTE*
WEP	*Wired Equivalent Privacy*
WFA	*Wi-Fi Alliance*
Wi-Fi	*Wireless Fidelity*
WLAN	*Wireless Local Area Network*
WLCP	*WLAN Control Plane*
WPA	*Wi-Fi Protected Access*
WRED	*Weighed Random Early Discard*
XML	*eXtensible Markup Language*

Introduction

The proliferation of mobile applications has increased the amount of data in the 4G mobile network. With the adoption of smartphones and broadband services, such as video streaming, cellular network resources are increasingly constrained.

Wi-Fi technology is ideally positioned to add capacity to the cellular network. It is necessary to improve the interworking between the 4G mobile network and the Wi-Fi network in order to offer a global and consistent broadband access to the end-user.

In addition to growing traffic, users expect unrestricted access to applications whether at home, in a business or on the road. For this reason, Wi-Fi technology, providing additional coverage, is an appropriate solution for roaming users.

The ability to exploit unlicensed frequency bands in addition to the spectrum allocated to cellular networks is of obvious appeal to network operators, who see Wi-Fi as another means of accessing the 4G mobile network.

Many mobile phones currently sold include both cellular and Wi-Fi radio access and are capable of simultaneously using both radios. This makes it possible to direct certain services to Wi-Fi access and others to the cellular radio access.

The various standardization bodies, IEEE (Institute of Electrical and Electronics Engineers), WFA (Wi-Fi Alliance) and 3GPP (3rd Generation Partnership Project), paved the way for the integration of Wi-Fi technology into the cellular network, allowing the mobile to access its services through Wi-Fi access.

I.1. 4G mobile network

I.1.1. *Network architecture*

The 4G mobile network, which is called EPS (Evolved Packet System), consists of an evolved packet core (EPC) and an evolved universal terrestrial radio access network (E-UTRAN) (Figure I.1).

The E-UTRAN access network provides the connection of the user equipment (UE). The core network EPC interconnects access networks, provides the interface to the packet data network (PDN) and provides mobile attachment and bearer establishment.

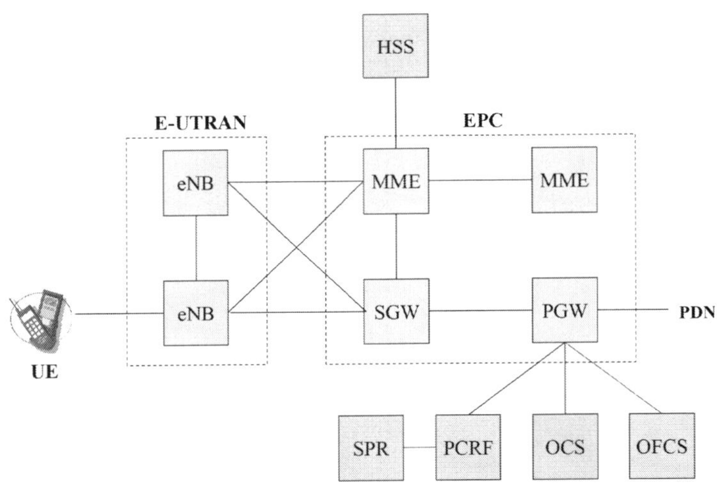

Figure I.1. *4G mobile network architecture*

The evolved node B station (eNB) compresses and encrypts traffic data on the radio interface, as well as encrypts and checks the integrity of signaling data exchanged with the mobile.

The mobility management entity (MME) allows mobile access to the EPS network and controls the establishment of bearers for the transmission of traffic data.

The SGW (Serving Gateway) entity is the anchor point for intra-system handover (mobility within the 4G network) and inter-system handover in packet-switched (PS) mode, requiring transfer of mobile traffic to a second- or third-generation mobile network.

The PGW (PDN Gateway) entity is the gateway router that connects the EPS network to the PDN. It provides the mobile with its configuration (IP address) and traffic information to the online charging system (OCS) for the prepaid and offline charging system (OFCS) for the postpaid.

The home subscriber server (HSS) is a database that stores data specific to each subscriber. The main stored data include subscriber identities, authentication parameters and service profile.

The policy charging and rules function (PCRF) provides the PGW entity with the rules to apply for the traffic (rate, quality of service, charging mode) when establishing the bearer. This information is stored in the subscription profile repository (SPR) when the subscription is created.

I.1.2. *Security architecture*

The mutual authentication between the mobile and the MME entity is based on the EPS-AKA (Authentication and Key Agreement) mechanism:

– the HSS entity provides the MME entity with the authentication vector (RAND, AUTN, RES, K_{ASME}) from the secret key Ki created during the subscription of the mobile;

– the MME entity provides the mobile with the random number (RAND) and the seal (AUTN) of the network;

– the mobile calculates the seals (AUTN, RES) and the key K_{ASME} from its key Ki stored in the universal subscriber identity module (USIM) of its universal integrated circuit card (UICC) and compares the seal (AUTN) received with that calculated;

– the mobile transmits its seal (RES) to the MME entity, which compares it to that received from the HSS entity;

– the K_{ASME} key is used to protect the signaling exchanged between the mobile and the MME entity as well as the control and traffic data on the radio interface.

I.1.3. *Bearer establishment*

The EPS network transports the mobile data stream (IP packets) transparently to the PGW entity that is routing the packets. The IP packet is transported in bearers built between the entities of the EPS network (Figure I.2).

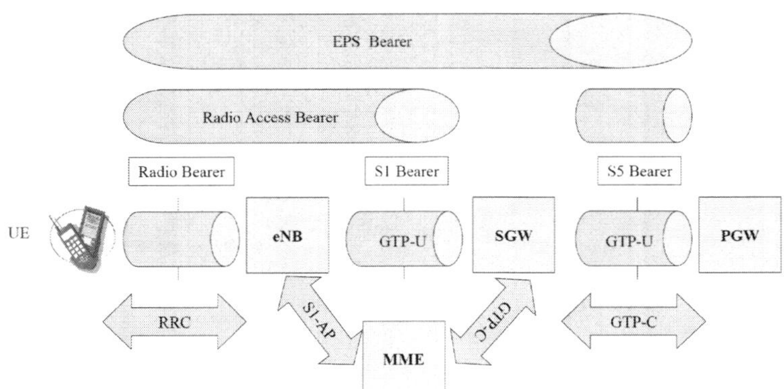

Figure I.2. *Bearer establishment*

The data radio bearer (DRB) is built between the mobile and the eNB entity. The RRC (Radio Resource Control) signaling, exchanged between the mobile and the eNB entity, is responsible for the construction of this bearer.

The S1 bearer is built between the eNB and SGW entities. The S1-AP signaling, exchanged between the eNB and MME entities, and the GTPv2 (GPRS Tunneling Protocol-Control) signaling, exchanged between the MME and SGW entities, are responsible for the construction of this bearer.

The S5 bearer is built between the SGW and PGW entities. The GTPv2-C signaling, exchanged between the SGW and PGW entities, is responsible for the construction of this bearer.

The connection of the radio bearer and the S1 bearer, carried out by the eNB entity, constitutes the EPS radio access bearer (E-RAB).

The connection of the E-RAB and S5 bearers, made by the SGW entity, constitutes the EPS bearer.

The S1 and S5 bearers are GTP-U (GPRS Tunneling Protocol User) tunnels, which allow the IP packet of the mobile to be transported in the IP packet of the bearer transmitted between the entities of the EPS network.

The PGW entity is the only entity in the EPS network that routes the mobile IP packet. The IP transport network that allows communication between the entities of the EPS network routes the IP packet that is the S1 or S5 bearer. The eNB and SGW entities do not perform routing. They only provide the connection between the bearers.

I.2. Wi-Fi network

I.2.1. *Network architecture*

The Wi-Fi (Wireless Fidelity) network consists of an access point (AP) that bridges the Wi-Fi radio interface with the Ethernet interface to the local area network (LAN) (Figure I.3).

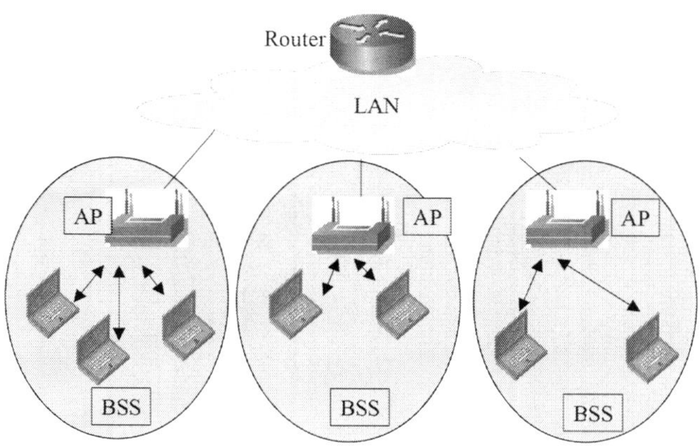

Figure I.3. *Wi-Fi network architecture*

The BSS (Basic Service Set) cell is the radio zone covered by the access point. The BSS identifier (BSSID) of the BSS cell is the MAC address of the access point.

Several BSS cells can be deployed to cover an area. The set of cells constitute an ESS (Extended Service Set) network. The ESS network is identified by the service set identifier (SSID).

Wi-Fi technology has defined the data link layer and physical layer of the radio interface (Figure I.4):

– the data link layer consists of two sub-layers, namely the LLC (Logical Link Control) sub-layer and the MAC (Medium Access Control) sub-layer;

– the physical layer has defined two sub-layers, namely the PLCP (Physical Layer Convergence Protocol) sub-layer and the PMD (Physical Medium Dependent) sub-layer.

Bridging consists of modifying the data link layer and the physical layer used on both sides of the access point.

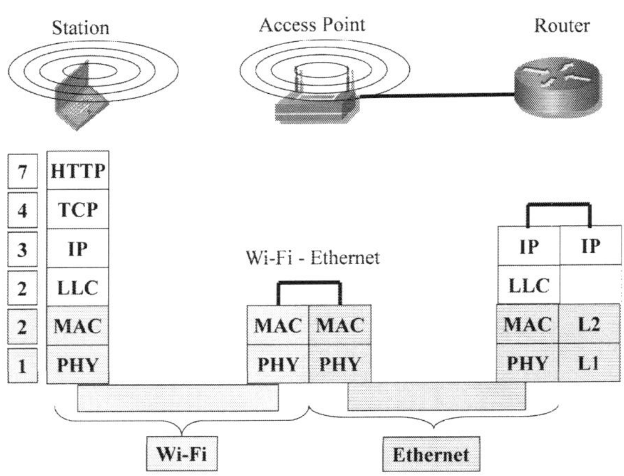

Figure I.4. *Protocol architecture*

The LLC sub-layer is not specific to Wi-Fi technology. It is also used for other data link layer protocols, such as the Ethernet MAC sub-layer. It indicates the nature of the encapsulated data, for example an IP packet.

The MAC sub-layer defines the procedure of access to the physical medium shared between the different mobiles of the cell. The CSMA/CA (Carrier Sense Multiple Access/Collision Avoidance) procedure solves the collision problems that occur when two mobiles simultaneously access the physical medium.

Particular MAC frames can be used for management functions (radio channel scanning, authentication, association) or transmission control (acknowledgment of received frames).

The PLCP sub-layer allows adaptation of the MAC sub-layer to the PMD sub-layer, providing signal-processing parameters for the receiver and indicating the bit rate of the frame.

The PMD sub-layer defines the characteristics of the radio transmission.

I.2.2. *Security architecture*

The 802.1x mechanism defines the mobile access control to the Wi-Fi network that is performed between the mobile and the RADIUS (Remote Authentication Dial-In User Service) server.

The 802.1x mechanism relies on EAP-Method (Extensible Authentication Protocol) authentication messages, for which several protocols are defined:

– EAP-CHAP (Challenge Handshake Authentication Protocol) protocol allows the authentication of the mobile by the RADIUS server, based on a password;

– EAP-TLS (Transport Layer Security) protocol allows mutual authentication of the RADIUS server and the mobile, based on certificates;

– EAP-TTLS (Tunneled Transport Layer Security) protocol allows mutual authentication of the RADIUS server based on certificate and of the mobile based on password.

Data protection on the radio interface introduces an extension of the MAC header:

– TKIP (Temporal Key Integrity Protocol) extension for the WPA (Wi-Fi Protected Access) mechanism based on RC4 (Rivest Cipher) algorithms for encryption and MICHAEL for integrity checking;

– CCMP (Counter-mode/CBC-MAC-Protocol) extension for the WPA2 mechanism based on the AES (Advanced Encryption Standard) algorithm for encryption and integrity checking.

I.2.3. *Physical layers*

The 802.11a interface defines the OFDM (Orthogonal Frequency Division Multiplexing) physical layer operating in the U-NII (Unlicensed-National Information Infrastructure) frequency band at 5 GHz.

The 802.11g interface defines the ERP (Extended Rate Physical) physical layer operating in the ISM (Industrial, Scientific and Medical) frequency band at 2.4 GHz.

The 802.11a/g interfaces have a bit rate of 6, 9, 12, 18, 24, 36, 48 or 54 Mbps depending on the modulation and coding scheme (MCS):

– the sub-carriers of the OFDM system are modulated in BPSK (Binary Phase Shift Keying), QPSK (Quadrature Phase Shift Keying), 16-QAM (Quadrature Amplitude Modulation) or 64-QAM;

– the binary convolutional coding (BCC) is used with a coding rate of 1/2, 2/3 or 3/4.

The 802.11n interface defines the HT (High Throughput) physical layer operating in the U-NII and ISM frequency bands at 5 and 2.4 GHz.

The 802.11n interface uses the OFDM system for which the modulation of the sub-carriers is the one defined for the 802.11a/g interfaces and introduces a new value (equal to 5/6) for the coding rate and a new error correction code LDPC (Low-Density Parity Check).

The 802.11n interface has a maximum rate of 600 Mbps obtained from two new features:

– the aggregation of two radio channels to obtain a bandwidth of 40 MHz;

– the spatial multiplexing SU-MIMO (Single User – Multiple Input Multiple Output) of two to four streams for a user.

The 802.11ac interface defines the VHT (Very High Throughput) physical layer operating only in the U-NII frequency band at 5 GHz.

The 802.11ac interface introduces new features to achieve a maximum rate of 6.9 Gbps:

– the aggregation of eight radio channels to obtain a bandwidth of 160 MHz;

– the spatial multiplexing SU-MIMO of two to eight streams for a user;

– the space multiplexing MU-MIMO (Multi-User – MIMO) supporting four users, with a maximum of four streams for each user, the total number of streams being limited to eight;

– the 256-QAM modulation.

I.3. Wi-Fi integration into the 4G mobile network

The integration of the Wi-Fi network into the 4G mobile network has an impact on the architecture of the EPC core network, which has several variants depending on the following characteristics:

– the Wi-Fi access is trusted or untrusted by the operator;

– the mobility is managed by the network or the mobile.

I.3.1. *Mutual authentication*

Mutual authentication is performed between the mobile and the AAA (Authentication, Authorization and Accounting) server. It uses the AKA mechanism adapted to the EAP-Method protocol:

– the HSS entity provides the AAA server with the authentication vector (RAND, AUTN, RES);

– the AAA server provides the mobile with the random number (RAND) and the seal (AUTN) of the network;

– the mobile calculates the seals (AUTN, RES) from its key Ki stored in the USIM module of its UICC card and compares the received seal (AUTN) with that calculated;

– the mobile transmits its seal (RES) to the AAA server, which compares it with that received from the HSS entity.

The EAP-AKA' protocol is an evolution of the EAP-AKA method, which concerns the key derivation mechanism.

I.3.2. *Architecture based on the S2a interface*

The architecture based on the S2a interface corresponds to trusted Wi-Fi access and network-based mobility.

The mobile stream travels through the Wi-Fi radio interface and tunnel S2a, built between the access point and the PGW entity, to access the PDN (Figure I.5).

The S2a interface supports several mechanisms for establishing the tunnel:

– the PMIPv6 (Proxy Mobile IP version 6) mechanism relies on the signaling provided by the mobility extension of the IPv6 header exchanged between the Wi-Fi access and the PGW entity and on the GRE (Generic Routing Encapsulation) tunnel for the mobile stream;

– the MIPv4 FA (Mobile IP version 4 Foreign Agent) mechanism is based on the MIPv4 signaling and the IP tunnel in IP for the mobile stream;

– the GTPv2 (GPRS Tunneling Protocol version 2) mechanism relies on the GTPv2-C signaling exchanged between the trusted Wi-Fi access and the PGW entity and on the GTP-U tunnel for the mobile stream.

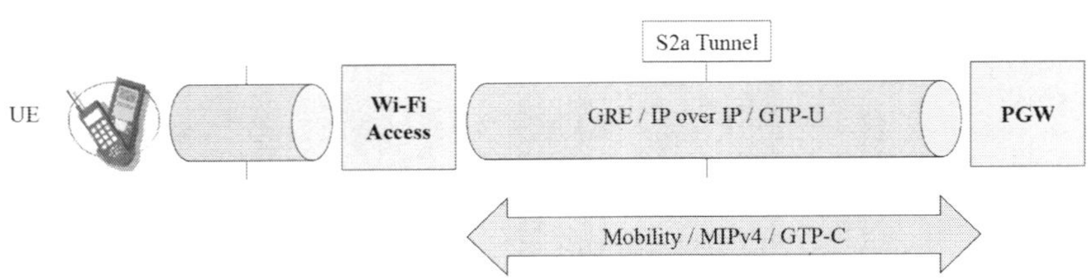

Figure I.5. *Session establishment – Architecture based on S2a interface*

I.3.3. *Architecture based on the S2b interface*

The architecture based on the S2b interface corresponds to untrusted Wi-Fi access and network-based mobility.

The mobile stream travels through the SWu tunnel, built between the mobile and the evolved packet data gateway (ePDG), and the S2b tunnel, built between the ePDG and PGW entities, to access the PDN (Figure I.6).

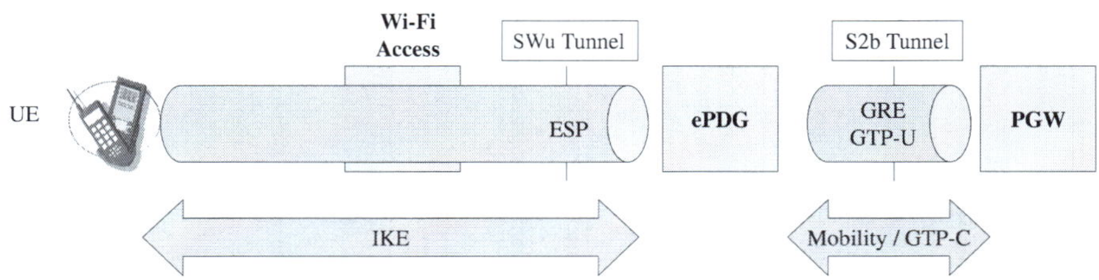

Figure I.6. *Session establishment – Architecture based on S2b interface*

The S2b interface supports the PMIPv6 or GTPv2 mechanism for tunnel establishment.

The SWu interface supports the IPSec (IP Security) mechanism, including IKEv2 (Internet Key Exchange version 2) signaling and the ESP (Encapsulating Security Payload) tunnel for the mobile stream.

I.3.4. *Architecture based on the S2c interface*

The architecture based on the S2c interface corresponds to trusted or untrusted Wi-Fi access and mobile-based mobility.

The mobile stream passes through the S2c tunnel built between the mobile and the PGW entity to access the PDN (Figure I.7).

In the case of untrusted Wi-Fi access, the S2c tunnel passes through the SWu tunnel built between the mobile and the ePDG entity (Figure I.7).

The S2c interface supports the DSMIPv6 (Dual-Stack Mobile IP version 6) mechanism for the establishment of the S2c tunnel built between the mobile and the PGW entity.

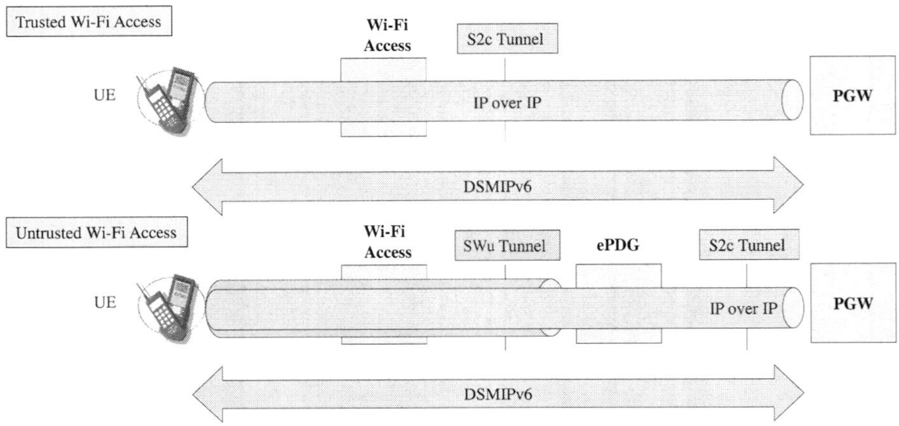

Figure I.7. *Session establishment – Architecture based on S2c interface*

In the case of trusted Wi-Fi access, this interface supports DSMIPv6 signaling and the IP tunnel in IP for the mobile stream.

In the case of trusted Wi-Fi access, the ESP tunnel, established between the mobile and the ePDG entity, protects the S2c interface.

I.3.5. *Network discovery and selection*

Mobile networks are becoming more and more heterogeneous. It is possible for a mobile to be covered simultaneously by different networks: traditional cellular networks, small cells integrating LTE and Wi-Fi accesses and stand-alone Wi-Fi access points. Given this variety, choosing the best network for a mobile is essential.

The access network discovery and selection function (ANDSF) allows network detection and selection between LTE and Wi-Fi accesses. The rules defined by the 4G mobile network operator are provided by the ANDSF server, which is an optional element of the EPC core network.

Hotspot 2.0 (HS2.0) is a working group of WFA. The target of the HS2.0 job is to facilitate the use of the Wi-Fi access point in a 4G mobile network. The HS2.0 certification program is called Passpoint.

The key features of version 1 are based on the 802.11u standard and include additions to the access point beacon and the ANQP (Access Network Query Protocol) server that provides rules defined by the Wi-Fi service operator.

Version 2 allows the mobile to identify the home operator and the partners that should be used when the home operator is not directly accessible.

I.4. Wi-Fi and LTE access aggregation

The integration of the Wi-Fi network to the 4G mobile network brings changes to the EPC core network, the anchor point being realized by the PGW entity. The aggregation of LTE and Wi-Fi channels is another approach that does not impact the structure of the EPC core network (Figure I.8).

LTE access operates in a licensed frequency band. The LTE Advanced and LTE Advanced Pro evolutions, respectively, defined an aggregation of 5 and 32 LTE channels. The eNB entity is the anchor point for channel aggregation.

LAA (Licensed Assisted Access) aggregation is an extension of LTE aggregation. The LTE transmission is performed on LTE and Wi-Fi frequency bands, between the mobile and the eNB entity, without an intermediate access point. The eNB entity is the anchor point for channel aggregation.

LWA (LTE-Wi-Fi Aggregation) uses LTE and Wi-Fi frequency bands. Transmission over the Wi-Fi radio channel is between the mobile and the access point in accordance with 802.11 standard. The eNB entity is the anchor point for channel aggregation.

MPTCP (Multi-Path Transmission Control Protocol) aggregation has the advantage of transmitting data using multiple paths without causing changes in existing infrastructures (4G mobile network, Wi-Fi network). The aggregation is performed by an MPTCP server.

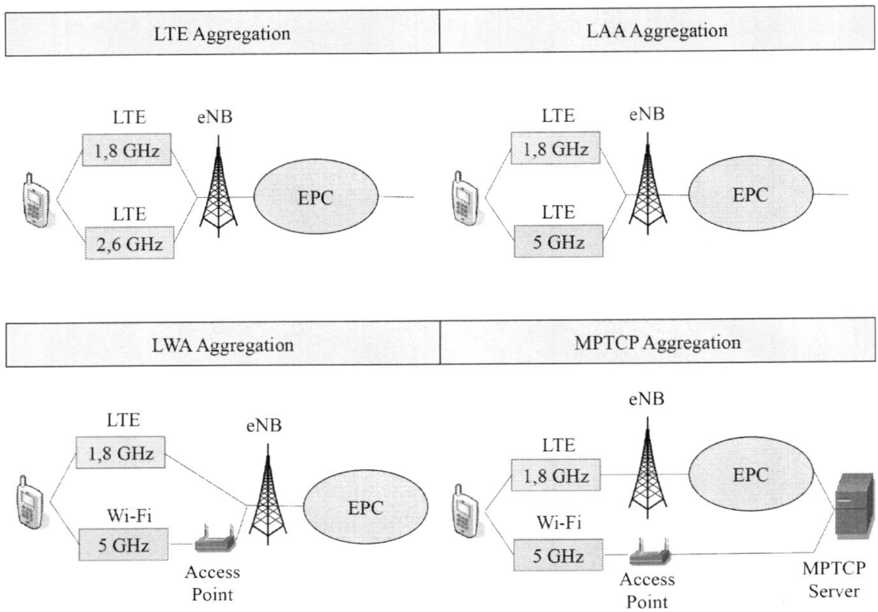

Figure I.8. *Wi-Fi and LTE access aggregation*

1

Architecture Based on Wi-Fi Access

1.1. Functional architecture

EPS (Evolved Packet System) is the name of the 4G mobile network. It consists of an evolved packet core (EPC) and an evolved universal terrestrial radio access network (E-UTRAN).

The E-UTRAN network presents the LTE (Long-Term Evolution) radio interface to the mobile.

Wi-Fi (Wireless Fidelity) interface is subsequently integrated into the EPS network and is a component of a set of technologies grouped under the term Non-3GPP Access.

Its introduction has an impact on the core network (EPC) architecture, which has several variants depending on the following characteristics:

- Wi-Fi access is trusted or untrusted by the operator;

- mobility is managed by the network or the mobile.

1.1.1. *Architecture based on the S2a interface*

The functional architecture based on the S2a interface corresponds to trusted Wi-Fi access and network-based mobility (Figure 1.1).

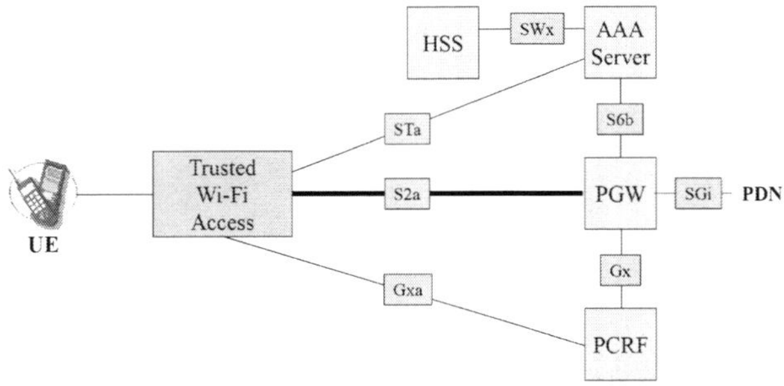

Figure 1.1. *Functional architecture based on the S2a interface*

The mobile stream travels through the Wi-Fi radio interface and the S2a tunnel to access the packet data network (PDN). The PGW (PDN Gateway) entity is an IP (Internet Protocol) router that acts as a gateway for the mobile stream.

The home subscriber server (HSS) and the AAA (Authentication, Authorization and Accounting) server provide the following functions:

– mutual authentication of the mobile and the AAA server via the interfaces SWx and STa. This authentication has the effect of opening Wi-Fi access to the mobile;

– transfer of the mobile profile comprising a list of access point names (APN) and the quality of service (QoS) level of the S2a tunnel and Wi-Fi interface, to the PGW entity, via the interface S6b, and to trusted Wi-Fi access, via the STa interface.

The policy charging and rules function (PCRF) also provides the traffic profile, including the QoS level of the S2a tunnel to the PGW entity, via the Gx interface, and to trusted Wi-Fi access via the Gxa interface.

The mobile profile is stored in the HSS entity for mounting the default bearers, and in this case, the presence of the PCRF is optional.

The presence of the PCRF entity is mandatory for the mounting of dedicated bearers on the initiative of an application function (AF), whose first example of implementation is the VoLTE (Voice over LTE) that provides telephone service.

The characteristics of the dedicated bearer of the IP packet containing the voice are only stored in the SPR (Subscriber Profile Repository) database associated with the PCRF entity.

Trusted WLAN access network (TWAN) includes the following features:

– WLAN AN: this feature includes Wi-Fi access points;

– TWAG (Trusted WLAN Access Gateway): this function terminates tunnel S2a;

– TWAP (Trusted WLAN AAA Proxy): this function terminates the STa interface.

The transparent connection mode provides a single connection to the PGW entity without mobility support between the LTE and Wi-Fi radio accesses. The IPv4 and/or IPv6 address of the mobile is provided by the TWAG function:

– in the case of a statefull configuration, the TWAG function acts as a DHCP (Dynamic Host Configuration Protocol) server;

– in the case of a stateless configuration, the TWAG function broadcasts the prefix of the IPv6 address.

The single-connection mode supports mobility between LTE and Wi-Fi accesses. This mode also supports non-seamless WLAN offload (NSWO), for which traffic is routed directly to the Internet network through TWAG function.

The multiple-connection mode supports NSWO and multiple-access PDN connectivity (MAPCON), for which the various connections to the PDN network pass through the LTE (e.g. telephone service) or Wi-Fi (e.g. Internet service) interfaces according to the policy of the operator. Mobility between LTE and Wi-Fi radio accesses is possible.

The connection on the Wi-Fi interface is established by the WLCP (WLAN Control Plane) protocol. The connection is identified by the MAC address of the mobile associated with a MAC address of the TWAG function.

For the single- or multiple-connection mode, the IPv4 and/or IPv6 address of the mobile is provided by the PGW.

The PGW entity shall allocate the downlink packets to different S2a bearers based on the TFT (Traffic Flow Template) packet filters set up during the establishment of the S2a bearer (Figure 1.2).

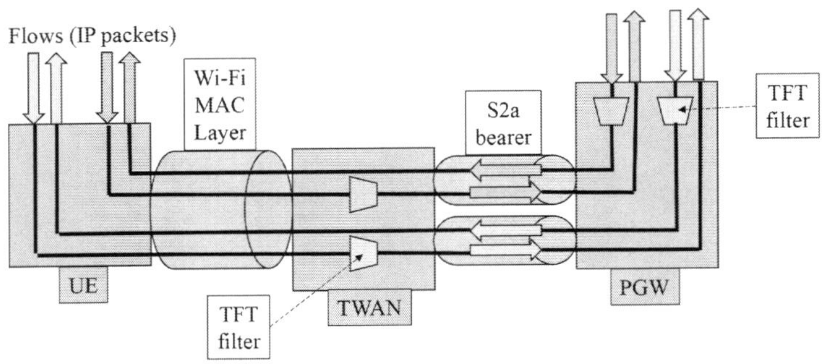

Figure 1.2. *Connection to the PDN network for architecture based on the S2a interface*

TWAN function of the trusted Wi-Fi access shall assign the uplink packets to different S2a bearers based on the TFT packet filters set up during the establishment of the S2a bearer (Figure 1.2).

1.1.2. *Architecture based on the S2b interface*

The functional architecture based on the S2b interface corresponds to untrusted Wi-Fi access and network-based mobility (Figure 1.3).

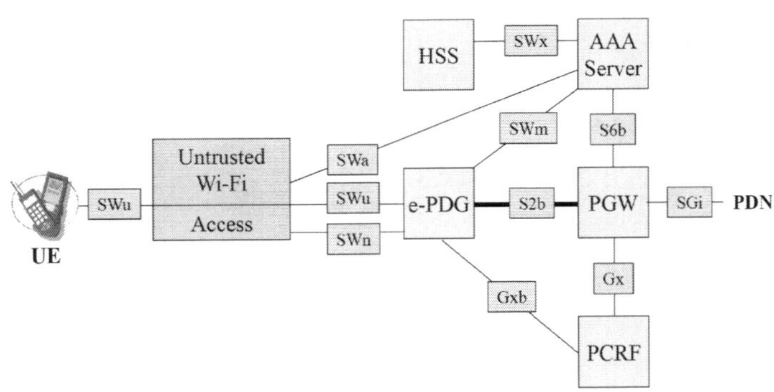

Figure 1.3. *Functional architecture based on the S2b interface*

The mobile stream passes through the SWu and S2b tunnels to access the PDN network via the PGW entity. The SWu tunnel is built between the mobile and the evolved packet data gateway (ePDG). The S2b tunnel is built between the ePDG and PGW entities.

The HSS entity and the AAA server provide the following functions:

– mutual authentication of the mobile and the AAA server, via the SWx and SWa interfaces. This authentication has the effect of opening Wi-Fi access to the mobile;

– mutual authentication related to the establishment of the SWu tunnel, via the SWx and SWm interfaces;

– transfer of the mobile profile comprising a list of access point names (APN) and the quality of service (QoS) level of the S2b tunnel, to the PGW entity via the interface S6b, to the ePDG entity via the SWm interface and to the untrusted Wi-Fi access via the SWa interface.

The PCRF entity provides the QoS level of the S2b tunnel to the PGW via the Gx interface and the ePDG via the Gxb interface.

The PCRF entity provides the QoS level of the SWu tunnel to the ePDG entity via the Gxb interface. In this case, the ePDG entity provides the QoS level to be applied on the Wi-Fi radio interface via the SWn interface.

The mobile must establish a SWu instance for each PDN connection.

When the mobile connects to the PDN network, a default bearer must be established on the S2b interface. This connection is maintained for the duration of the connection.

Dedicated bearers can be built for the same PDN connection, based on the rules provided by the PCRF.

An SWu instance transports the packets of all the S2b bearers for the same connection to the PDN network between the mobile and the ePDG entity.

The ePDG entity shall release the SWu instance when the S2b default bearer of the associated connection to the PDN network is released.

Two IPv4 and/or IPv6 addresses are assigned to the mobile:

– an address for the SWu tunnel built between the mobile and the ePDG entity, provided by the untrusted Wi-Fi access;

– an address for the flow transiting in this tunnel, provided by the PGW entity.

The connection to the PDN network is described in Figure 1.4.

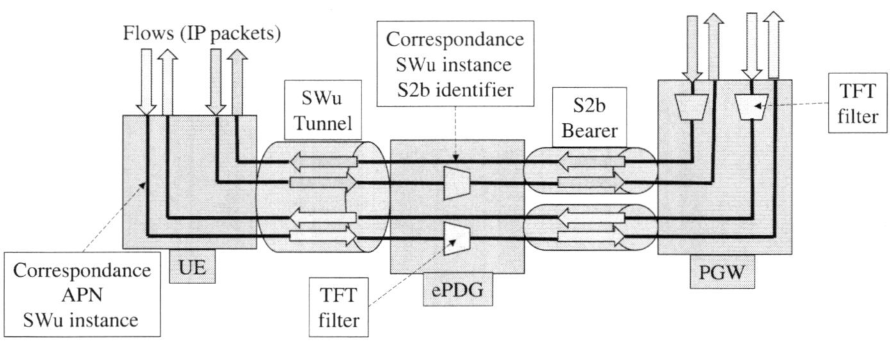

Figure 1.4. *Connection to the PDN network for architecture based on S2b interface*

The PGW entity must allocate the downlink packets to different S2b bearers according to the TFT packet filters set up during the establishment of the S2b bearer.

The ePDG entity must assign the downlink packets to the SWu instance based on the correspondence between the SWu instance and the identifier of the S2b bearer.

The mobile must assign the uplink packets to the SWu instance based on the correspondence between the APN identifier of the PDN connection and the SWu instance.

The ePDG entity must allocate the uplink packets to different S2b bearers according to the TFT packet filters set up during the establishment of the S2b bearer.

1.1.3. *Architecture based on the S2c interface*

The functional architecture based on the S2c interface corresponds to a mobility based on the mobile. The functional architecture is depicted in Figure 1.5 for trusted Wi-Fi access and Figure 1.6 for untrusted Wi-Fi access.

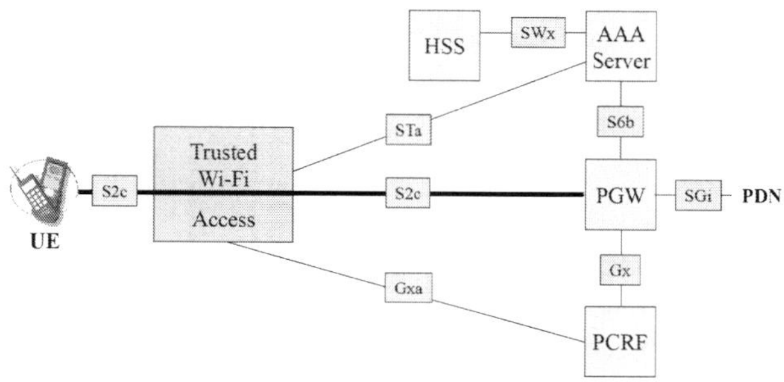

Figure 1.5. *Functional architecture based on S2c interface Trusted Wi-Fi access*

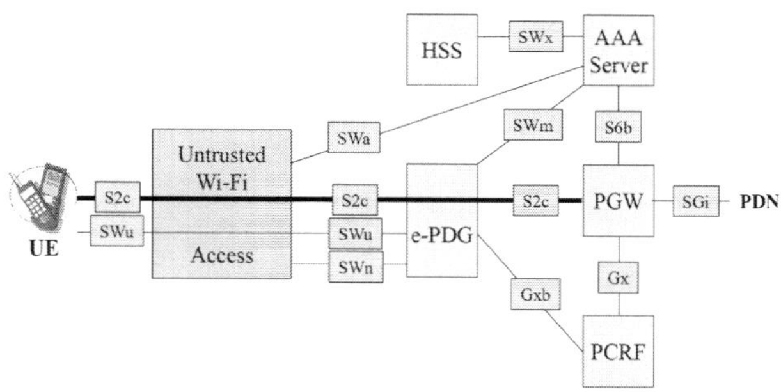

Figure 1.6. *Functional architecture based on S2c interface Untrusted Wi-Fi access*

The mobile stream passes through the S2c tunnel built between the mobile and the PGW entity to access the PDN data network.

In the case of untrusted Wi-Fi access, the S2c tunnel passes through the SWu tunnel built between the mobile and the ePDG entity.

1.2. Tunnel establishment

1.2.1. *Architecture based on the S2a interface*

The S2a interface is the point of reference between the PGW entity and the trusted Wi-Fi access. This interface supports several mechanisms for the establishment of the S2a tunnel.

The construction of S2a tunnel requires the selection of the PGW entity by Wi-Fi access, from information provided by the AAA server during authentication.

This information can be the IP address of the PGW entity, the full qualified domain name (FQDN) or the APN. Trusted Wi-Fi access retrieves the IP address of the PGW entity by performing DNS (Domain Name System) resolution on the FQDN or APN.

1.2.1.1. *PMIPv6 mechanism*

The PMIPv6 (Proxy Mobile IP version 6) mechanism relies on the signaling provided by the mobility extension of the IPv6 header exchanged between Wi-Fi access and the PGW entity (Figure 1.7) and on the GRE (Generic Routing Encapsulation) tunnel of the mobile stream (Figure 1.8).

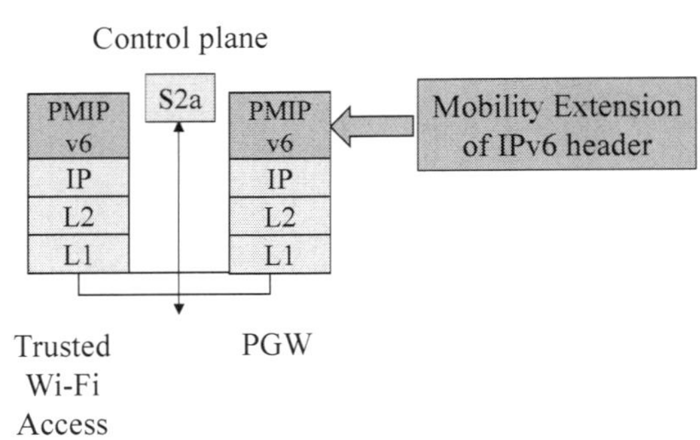

Figure 1.7. *Protocol architecture based on S2a interface Control plane for PMIPv6 mechanism*

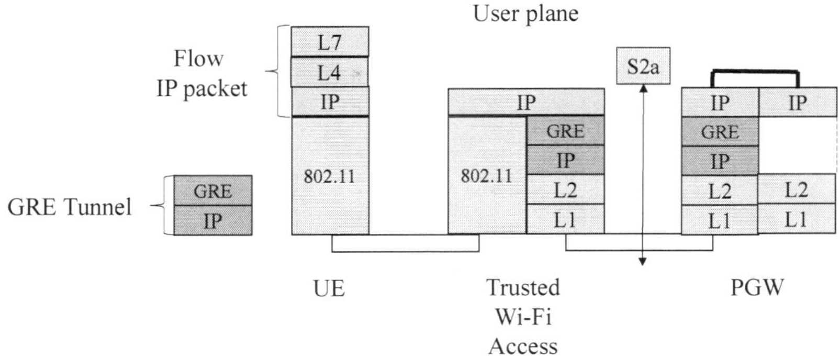

Figure 1.8. *Protocol architecture based on S2a interface User plane for PMIPv6 mechanism*

The MIPv6 mechanism requires functionality in the IPv6 stack of a mobile node. The exchange of signaling messages between the mobile node and the home network agent makes it possible to create and maintain a correspondence between its address in the home network and the foreign network.

Network-based mobility supports the mobility of IPv6 nodes without mobile involvement by extending MIPv6 signaling between the TWAG function and the PGW entity.

This approach to support mobility does not require the mobile node to be involved in the exchange of signaling messages. The PMIPv6 protocol is an extension of the MIPv6 protocol.

A mobile node can operate in an IPv4, IPv6 or IPv4/IPv6 environment. The PMIPv6 protocol independently supports the mobility of the IPv4 address and the transport of IP packets in an IPv4 network.

1.2.1.2. *MIPv4 mechanism*

The MIPv4 FA (Mobile IP version 4 Foreign Agent) mechanism is based on MIPv4 signaling (Figure 1.9) and the IP in the IP tunnel of the mobile stream (Figure 1.10).

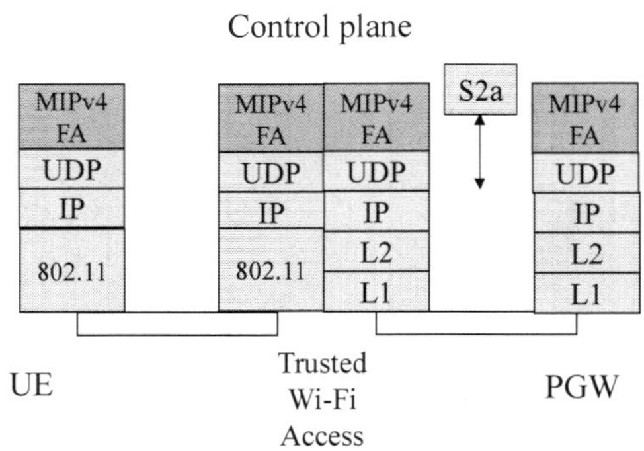

Figure 1.9. *Protocol architecture based on S2a interface Control plane for MIPv4 FA mechanism*

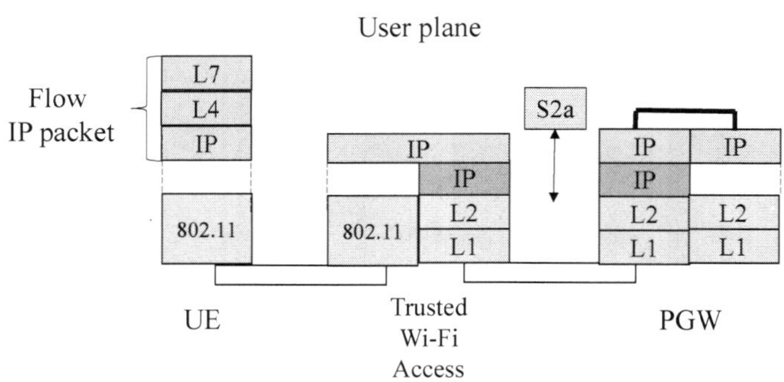

Figure 1.10. *Protocol architecture based on S2a interface User plane for MIPv4 FA mechanism*

MIPv4 signaling is exchanged, on the one hand, between the mobile and trusted Wi-Fi access and, on the other hand, between the trusted Wi-Fi access and the PGW entity.

The MIPv4 protocol allows Wi-Fi access, playing the role of a foreign agent, to assign the mobile an IPv4 address in a foreign network.

The MIPv4 protocol makes it possible to register with the PGW entity, which plays the role of a home agent, the correspondence between the mobile IPv4 address in the home network, provided by the PGW entity, and the IPv4 address in the foreign network.

1.2.1.3. *GTPv2 mechanism*

The GTPv2 (GPRS Tunneling Protocol version 2) mechanism is based on the GTPv2-C (Control) signaling exchanged between the trusted Wi-Fi access and the PGW entity (Figure 1.11) and on the GTP-U (User) tunnel of the mobile flow (Figure 1.12).

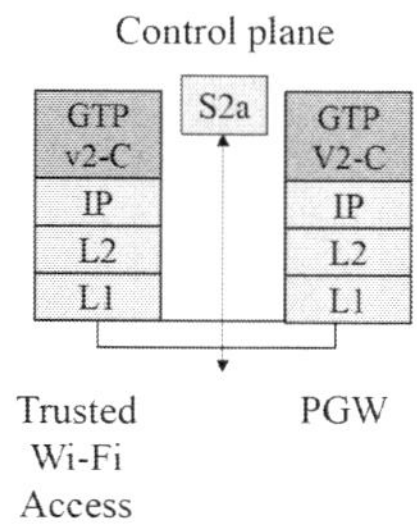

Figure 1.11. *Protocol architecture based on S2a interface Control plane for GTPv2 mechanism*

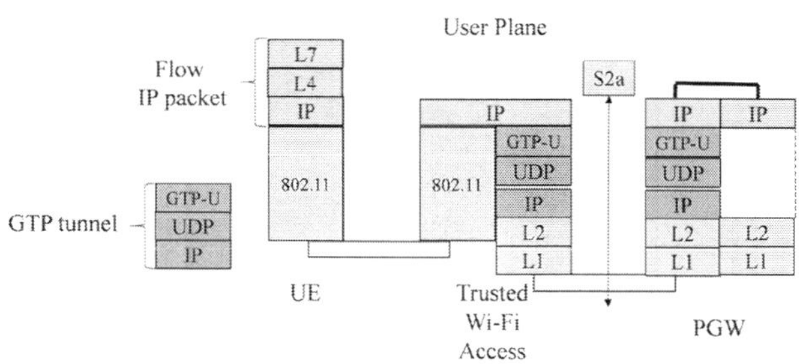

Figure 1.12. *Protocol architecture based on S2a interface User plane for GTPv2 mechanism*

The GTPv2-C protocol allows the activation or deactivation of a session as well as the creation, modification or release of GTP-U bearers.

The PMIPv6 and GTPv2 mechanisms can transport IPv4 or IPv6 streams in IPv4 or IPv6 tunnels. The MIPv4 mechanism allows the transport of only IPv4 streams in IPv4 tunnels.

1.2.2. *Architecture based on the S2b interface*

The S2b interface is the point of reference between the PGW and ePDG entities. This interface supports the PMIPv6 (Figures 1.13 and 1.14) or GTPv2 mechanism for the establishment of the S2b tunnel.

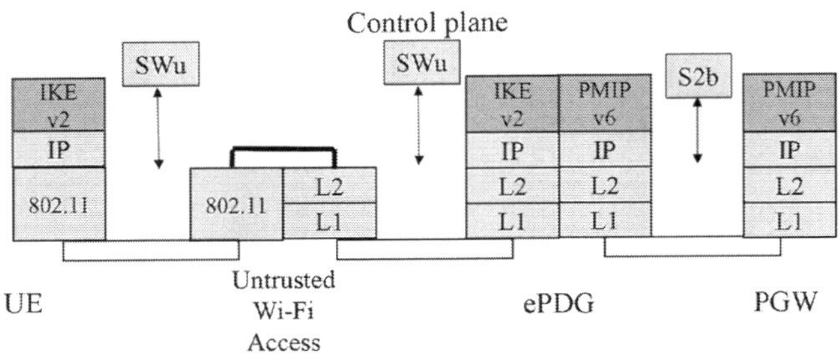

Figure 1.13. *Protocol architecture based on S2b interface Control plane for PMIPv6 mechanism*

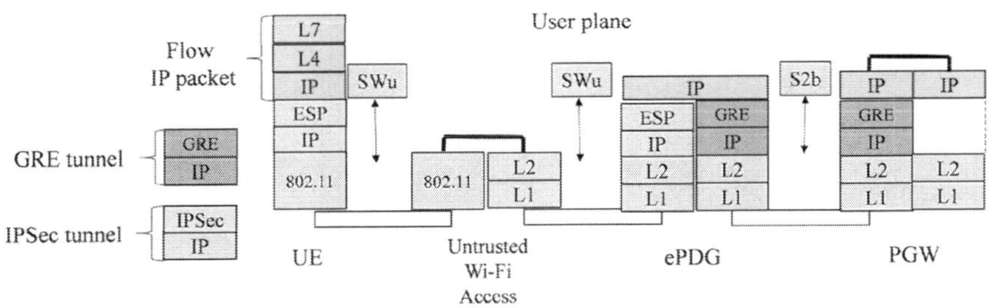

Figure 1.14. *Protocol architecture based on S2b interface User plane for PMIPv6 mechanism*

The SWu interface is the point of reference between the ePDG entity and the mobile. This interface supports the IPSec (IP Security) mechanism, including IKEv2 (Internet Key Exchange version 2) signaling (Figure 1.13) and the ESP (Encapsulating Security Payload) tunnel of the mobile stream (Figure 1.14).

The construction of the SWu tunnel requires the retrieval of the IP address of the ePDG entity by the mobile. This IP address can be configured in the mobile by various means.

The mobile can also perform a DNS resolution on the FQDN of the ePDG entity. The mobile automatically builds the FQDN from the identity of the operator contained in its international mobile subscriber identity (IMSI) or from the tracking area identifier (TAI), where the mobile is located.

The construction of the S2b tunnel requires the selection of the PGW entity by the ePDG entity, from information provided by the AAA server during the authentication for the establishment of the SWu tunnel.

1.2.3. *Architecture based on the S2c interface*

The S2c interface is the point of reference between the PGW entity and the mobile. This interface supports the DSMIPv6 (Dual-Stack Mobile IP version 6) mechanism for the establishment of the S2c tunnel built between the mobile and the PGW entity.

In the case of trusted Wi-Fi access, this interface supports DSMIPv6 signaling (Figure 1.15) and IP in IP tunnel (Figure 1.16) of the mobile stream.

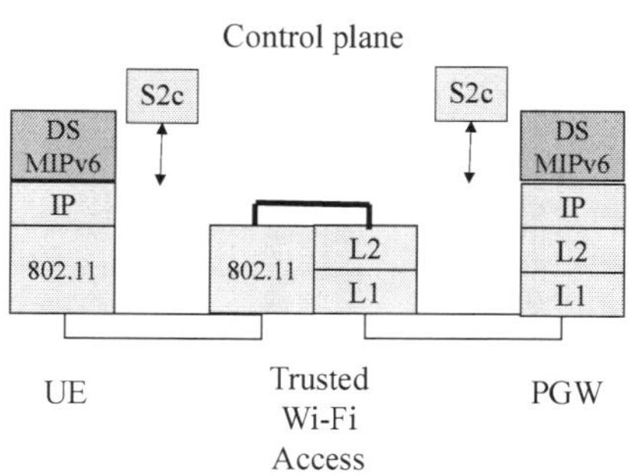

Figure 1.15. *Protocol architecture based on S2c interface Control plane for trusted Wi-Fi access*

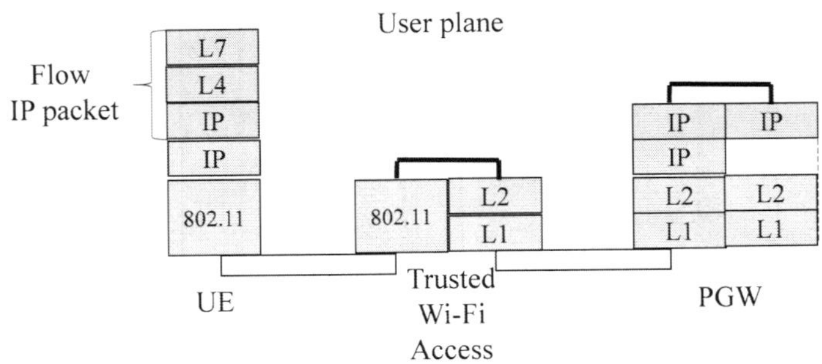

Figure 1.16. *Protocol architecture based on S2c interface User plane for trusted Wi-Fi access*

In the case of untrusted Wi-Fi access, the IPSec tunnel established between the mobile and the ePDG entity protects the S2c interface.

The MIPv6 protocol allows IPv6 mobile nodes to move while maintaining accessibility and ongoing sessions.

The DSMIPv6 protocol prevents the IPv4/IPv6 dual-stack mobile from running both MIPv4 and MIPv6 mobility protocols simultaneously.

The DSMIPv6 protocol also takes into account the case where the mobile moves in a private IPv4 network. The mobile node must be able to communicate with the PGW entity, which acts as a home agent, through a NAT (Network Address Translation) device.

In the case of untrusted Wi-Fi access, the S2c tunnel is established from the IP address of the PGW provided by the AAA server during the authentication for the establishment of the SWu tunnel.

The mobile can also retrieve the IP address of the PGW entity by querying a DHCP (Dynamic Host Configuration Protocol) server or by performing DNS resolution on the FQDN of the PGW.

1.3. DIAMETER protocol

The DIAMETER protocol is used to perform authentication, authorization and accounting functions.

The authentication function makes it possible to control the access of the mobile to the 4G mobile network from a stored secret, on the one hand, in the universal subscriber identity module (USIM) of the universal integrated circuit card (UICC) of the mobile and, on the other hand, in the HSS entity.

The authorization function retrieves the service and traffic profile of the mobile stored in the HSS and SPR databases.

The accounting function allows generation of events from the PGW entity to the charging entities for the prepaid or postpaid service.

1.3.1. *AAA server interfaces*

The DIAMETER protocol is supported on the interfaces between, on the one hand, the AAA server and, on the other hand (Figure 1.17):

– trusted Wi-Fi access via the STa interface;

– untrusted Wi-Fi access via the SWa interface;

– PGW entity via the S6b interface;

– ePDG entity via the SWm interface;

– HSS entity via the SWx interface.

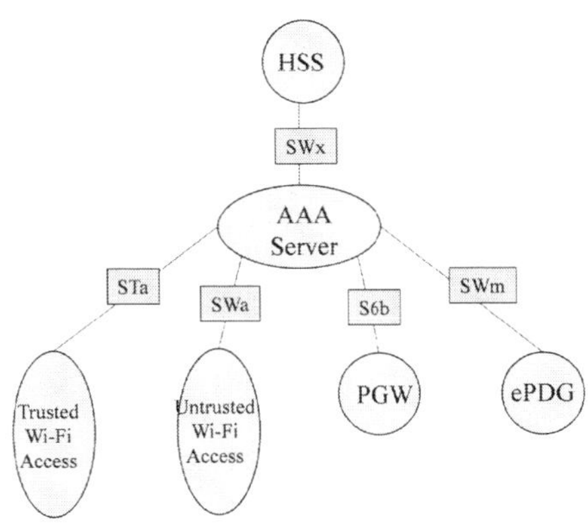

Figure 1.17. *AAA server interfaces using the DIAMETER protocol*

The SWx interface is used by the AAA server to retrieve the authentication data; the subscriber profile and the parameters for the PMIPv6, MIPv4 FA, GTPv2 and DSMIPv6 mechanisms.

The SWx interface is used to register the address of the PGW and the AAA server in the HSS when establishing tunnel S2a, S2b or S2c.

The SWx interface is used by the HSS entity for updating the mobile profile and for detaching it.

Table 1.1 summarizes the DIAMETER messages exchanged on the SWx interface.

Messages	Comments
Multimedia-Authentication-Request (MAR)	AAA server request to retrieve authentication data
Multimedia-Authentication-Answer (MAA)	HSS entity response containing authentication data
Server-Assignment-Request (SAR)	AAA server request to register the PGW entity and retrieve the mobile profile
Server-Assignment-Answer (SAA)	HSS entity response containing mobile profile
Registration-Termination-Request (RTR)	HSS server request for mobile detachment
Registration-Termination-Answer (RTA)	AAA server response to RTR request
Push-Profile-Request (PPR)	HSS entity request for mobile profile update
Push-Profile-Answer (PPA)	AAA server response to PPR request

Table 1.1. *DIAMETER messages on the SWx interface*

The STa and SWa interfaces share the same authentication procedure. During the authentication phase, the AAA server decides whether Wi-Fi access is trusted or untrusted and communicates the decision to the Wi-Fi access point.

The STa and SWa interfaces are used to carry information relating to the PMIPv6, MIPv4 FA (only in the case of the STa interface), GTPv2 and DSMIPv6 mechanisms.

The STa and SWa interfaces are used for detaching the mobile, the procedure being at the initiative of the Wi-Fi access or the AAA server.

The STa and SWa interfaces are used to renew mobile authentication. The procedure is initiated by the AAA server in the event that the subscriber's profile stored in the HSS entity is changed, or at the initiative of the Wi-Fi access that wants to verify that the subscriber's profile is not modified.

Table 1.2 summarizes the DIAMETER messages exchanged on the STa and SWa interfaces.

Messages	Comments
Authenticate and Authorize Request (AAR)	Wi-Fi access request to register and retrieve the mobile profile
Authenticate and Authorize Answer (AAA)	AAA server response containing mobile profile
Re-Auth-Request (RAR)	AAA server request for mobile authentication renewal
Re-Auth-Answer (RAA)	Response from Wi-Fi access to RAR request
Session Termination Request (STR)	Wi-Fi access request for ending the mobile session
Session Termination Answer (STA)	AAA server response to STR request
Abort-Session-Request (ASR)	AAA server request for termination of mobile session
Abort-Session-Answer (ASA)	Response from Wi-Fi access to ASR request
Diameter-EAP-Request (DER)	Wi-Fi access request used for the EAP-AKA authentication procedure
Diameter-EAP-Answer (DEA)	AAA server response used for the EAP-AKA authentication procedure

Table 1.2. *DIAMETER messages on the STa and SWa interfaces*

The S6b interface is used by the PGW entity to communicate to the AAA server its address when the tunnel S2a, S2b or S2c is established.

The S6b interface is used by the PGW entity to retrieve the subscriber's profile and the PMIPv6 and GTPv2 mechanism information.

The S6b interface is used by the PGW entity to retrieve mobile authentication data for the DSMIPv6 mechanism. The authentication data is used to control the establishment of the IPSec mechanism to protect the DSMIPv6 signaling exchanged between the mobile and the PGW entity.

The S6b interface is used for terminating the mobile session, the procedure being initiated by the PGW entity or the AAA server.

Table 1.3 summarizes the DIAMETER messages exchanged on the S6b interface.

Messages	Comments
Authenticate and Authorize Request (AAR)	PGW entity request to register and retrieve the mobile profile
Authenticate and Authorize Answer (AAA)	AAA server response containing mobile profile
Re-Auth-Request (RAR)	AAA server request for mobile authentication renewal
Re-Auth-Answer (RAA)	PGW response to RAR request
Session Termination Request (STR)	PGW request for termination of mobile session
Session Termination Answer (STA)	AAA server response to STR request
Abort-Session-Request (ASR)	AAA server request for termination of mobile session
Abort-Session-Answer (ASA)	PGW response to ASR request
Diameter-EAP-Request (DER)	Request of the PGW entity used for the EAP-AKA authentication procedure for the DSMIPv6 mechanism
Diameter-EAP-Answer (DEA)	AAA server response used for the EAP-AKA authentication procedure

Table 1.3. *DIAMETER messages on the S6b interface*

The SWm interface is used for the mutual authentication procedure of the mobile and the AAA server, which is implemented during the establishment of the SWu tunnel.

The SWm interface is used by the ePDG entity to retrieve the subscriber's profile and the PMIPv6 and GTPv2 mechanism information.

The SWm interface can also be used to transmit to the ePDG entity, the IP address or the FQDN of the PGW entity.

The SWm interface is used for terminating the mobile session, the procedure being initiated by the ePDG entity or the AAA server.

Table 1.4 summarizes the DIAMETER messages exchanged on the SWm interface.

Messages	Comments
Authenticate and Authorize Request (AAR)	Request from the ePDG entity to register itself and retrieve the mobile profile
Authenticate and Authorize Answer (AAA)	AAA server response containing mobile profile
Re-Auth-Request (RAR)	AAA server request for mobile authentication renewal
Re-Auth-Answer (RAA)	Response of the ePDG entity to the RAR request
Session Termination Request (STR)	Request from ePDG entity for termination of mobile session
Session Termination Answer (STA)	AAA server response to STR request
Abort-Session-Request (ASR)	AAA server request for termination of mobile session
Abort-Session-Answer (ASA)	Response of the ePDG entity to the ASR request
Diameter-EAP-Request (DER)	Request of the ePDG entity used for the EAP-AKA authentication procedure for the DSMIPv6 mechanism
Diameter-EAP-Answer (DEA)	AAA server response used for the EAP-AKA authentication procedure

Table 1.4. *DIAMETER messages on the SWm interface*

1.3.2. *PCRF interfaces*

The DIAMETER protocol is also supported on the interfaces between, on the one hand, the PCRF entity and, on the other hand (Figure 1.18):

– PGW entity via the Gx interface;

– trusted Wi-Fi access via the Gxa interface;

– ePDG entity via the Gxb interface.

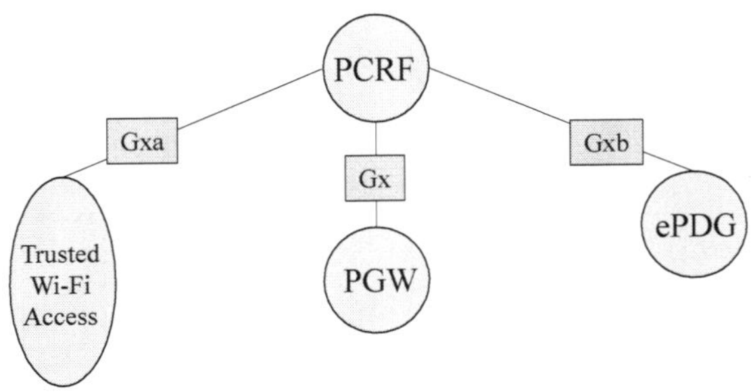

Figure 1.18. *PCRF interfaces using the DIAMETER protocol*

The Gx, Gxa and Gxb interfaces make it possible to request the PCRF entity to:

– retrieve the rules to apply to the default bearer created by the EPS network;

– inform the PCRF entity of the termination of the session on the EPS network.

The Gx, Gxa and Gxb interfaces allow the PCRF entity to provide the rules to be applied for the dedicated bearer.

Table 1.5 summarizes the DIAMETER messages exchanged on the Gx, Gxa and Gxb interfaces.

Messages	Comments
Credit-Control-Request (CCR)	Request from PGW, ePDG or trusted Wi-Fi entities to retrieve the mobile profile
Credit-Control-Answer (CCA)	PCRF response containing the mobile profile
Re-Auth-Request (RAR)	Request from the PCRF entity containing the mobile profile
Re-Auth-Answer (RAA)	Response of PGW, ePDG or trusted Wi-Fi access to the RAR request

Table 1.5. *DIAMETER messages on the Gx, Gxa and Gxb interfaces*

2

MAC Layer

2.1. Frame structure

2.1.1. *Frame header*

The MAC (Medium Access Control) header, described in Figure 2.1, encapsulates an LLC (Logical Link Control) frame whose size is less than or equal to 2,304 bytes.

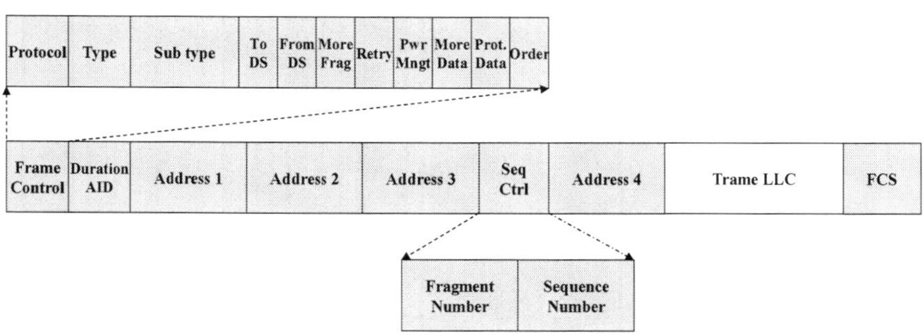

Figure 2.1. *MAC header structure*

Frame Control: this field consists of a sequence of several subfields:

– Protocol Version: this subfield is coded on two bits and takes the value 00;

– Type and Subtype: the subfields are coded, respectively, on two and four bits. They identify the function of the frame. There are three types of frames, namely the traffic frame, the control frame and the management frame. For each type of frame, subtypes are defined;

– To DS and From DS: these two subfields are coded on one bit. They indicate the direction of transmission of the frame (Table 2.1);

– More Fragments: this subfield is coded on a bit. It takes the value of ONE for traffic or management frames, if other fragments follow;

– Retry: this subfield is coded on a bit. It takes the value of ONE to signal the retransmission of a frame;

– Power Management: this subfield is coded on a bit. It takes the value of ONE when the station signals the switch to standby state;

– More Data: this subfield is coded on a bit. It takes the value of ONE when the access point signals to the terminal that frames are stored in the buffer;

– Protected Frame: this subfield is coded on a bit. It takes the value of ONE when the frame payload is secured by the WPA1 (Wi-Fi Protected Access) or WPA2 mechanism;

– Order: this subfield is coded on a bit. It takes the value of ONE to indicate that the frame is transmitted as part of an ordered service.

To DS	From DS	Meaning
0	0	All control and management frames Traffic frames in ad hoc mode
0	1	Traffic frames to the local area network (LAN)
1	0	Traffic frames from the LAN
1	1	Traffic frames exchanged between access points

Table 2.1. *To DS and From DS subfield values*

Duration/AID: this field is coded on 16 bits:

– Duration indicates the time during which the radio resource is immobilized;

– AID (Association Identifier) indicates the name of an association identifier in the case of the transmission of a PS (Power Save)-POLL control frame.

Address: there are four address fields, each of which is six bytes long. The construction rule is identical to that of an Ethernet MAC address. These fields indicate the basic service set identifier (BSSID), source address (SA), destination address (DA), transmitter address (TA) and receiver address (RA).

To DS	From DS	Address 1	Address 2	Address 3	Address 4
0	0	DA	SA	BSSID	
1	0	BSSID	SA	DA	
0	1	DA	BSSID	SA	
1	1	RA	TA	DA	SA

Table 2.2. *Meaning of Address fields*

Sequence Control: this field contains two subfields:

– Sequence Number: this subfield is coded on 12 bits. It indicates the number of the frame modulo-4096;

– Fragment Number: this subfield is coded on four bits. It indicates the number of the fragment in the frame. The value is equal to ZERO for the first fragment. All fragments of the same frame have the same value of the frame number.

FCS (Frame Check Sequence): this field is coded on 32 bits. It contains the cyclic redundancy code for error detection.

2.1.2. *Structure of control frames*

The Type subfield is set to 01 for control frames.

The RTS (Request To Send) frame is transmitted by the station to request the access point to access to the radio resource. The RA field contains the MAC address of the access point and the TA field of the station (Figure 2.2). The Subtype subfield is set to 1011.

The CTS (Clear To Send) frame is transmitted by the access point to allow the station to access the radio resource. The RA field contains the MAC address of the station (Figure 2.2). The Subtype subfield is set to 1100.

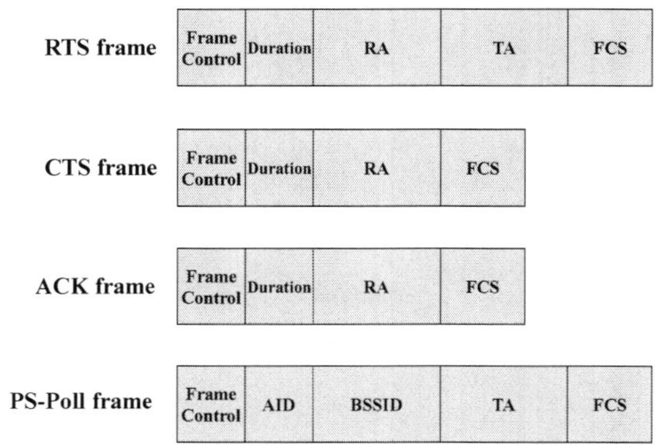

Figure 2.2. *Structure of control frames*

The ACK (Acknowledgment) frame is transmitted to acknowledge the received frame. This can be a traffic frame, a management frame or the PS-Poll control frame. The RA field copies the MAC address contained in the Address 2 field of the received frame (Figure 2.2). The Subtype subfield is set to 1101.

The PS-POLL frame is sent by the station to warn the access point that it has left sleep mode. The BSSID field contains the MAC address of the access point and the TA field of the station. The AID field is an identifier assigned to the station during the association phase (Figure 2.2). The Subtype subfield is set to 1010.

2.1.3. *Structure of management frames*

The Type subfield is set to 00 for management frames.

The BEACON management frame is a beacon channel that broadcasts information on the network. It contains mandatory fields and optional fields (Figure 2.3). The Subtype subfield is set to 1000.

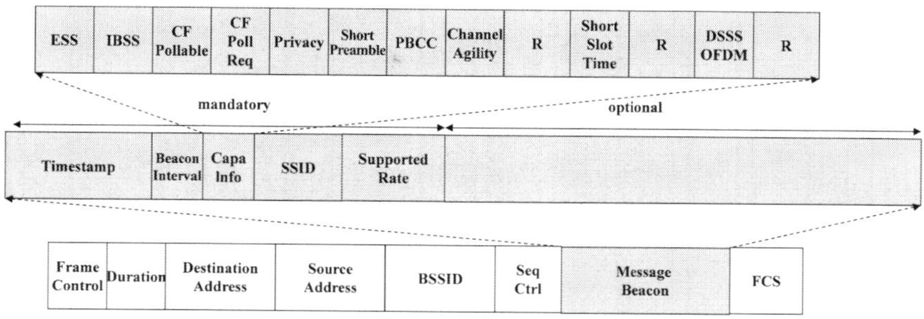

Figure 2.3. *Structure of the BEACON management frame*

Timestamp: this field is coded on 64 bits. It contains the timestamp of the frame.

Beacon Interval: this field is coded on 16 bits. It indicates the frequency of emission of the beacon channel.

Capability Information: this field is coded on 16 bits. It contains the characteristics of the access point:

– the type of network architecture (ESS, IBSS);

– the implementation of the security (Privacy);

– the use of a short preamble for the 802.11g radio interface;

– the use of a short slot time for the 802.11g radio interface;

– the use of the DSSS-OFDM physical layer for the 802.11g radio interface.

SSID (Service Set Identifier): this field has a variable length less than or equal to 34 bytes. It provides the identifier of the ESS (Extended Service Set) network.

Supported Rates: this field is composed of several information elements. Each element has a variable length less than or equal to 10 bytes and specifies the rates supported by the access point.

The PROBE REQUEST management frame is used by the station to request the characteristics of the radio interface of the access point. The PROBE REQUEST frame is a broadcast frame. The Subtype subfield is set to 0100.

When the station has sent the PROBE REQUEST frame, it will arm a timer. If there is no response before expiration, then the station repeats the process on another radio channel.

The access point provides its characteristics in the PROBE RESPONSE management frame when the value of the SSID contained in the PROBE REQUEST frame corresponds to that of the access point. The PROBE RESPONSE management frame is transmitted in unicast. The Subtype subfield is set to 0101.

The AUTHENTICATION management frame is used for the authentication of the station (Figure 2.4).

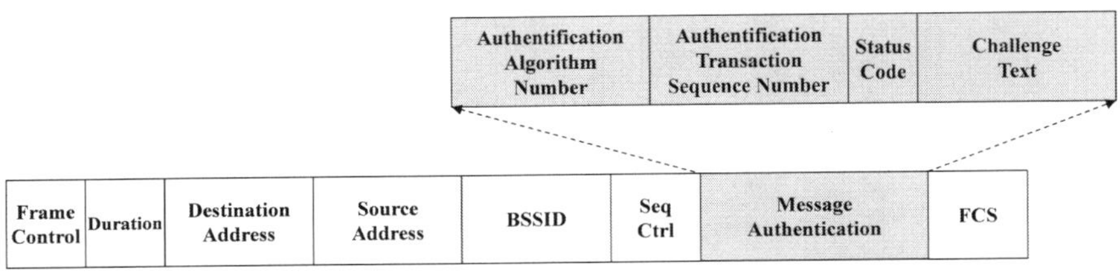

Figure 2.4. *Structure of the AUTHENTICATION management frame*

Authentication Algorithm Number: this field is coded on 16 bits. It identifies the authentication mode. The following two modes are defined:

– OSA (Open System Authentication): this mode corresponds to open access to the network. This mode is used for the WPA1 and WPA2 mechanisms;

– SKA (Shared Key Authentication): this mode requires the station to send a seal to access the network. This mode is used for the WEP (Wired Equivalent Privacy) mechanism.

Authentication Transaction Sequence Number: this field is coded on 16 bits. It contains the number of the authentication sequence.

Status Code: this field is coded on 16 bits. It indicates whether the operation was successful or not.

Challenge Code: this field, used for the WEP mechanism, has a variable size less than or equal to 255 bytes. It contains a string of bits, emitted in clear by the access point and then encrypted by the station.

The association phase is implemented from four management frames, namely ASSOCIATION REQUEST, ASSOCIATION RESPONSE, REASSOCIATION REQUEST and REASSOCIATION RESPONSE. These frames introduce new fields (Figure 2.5).

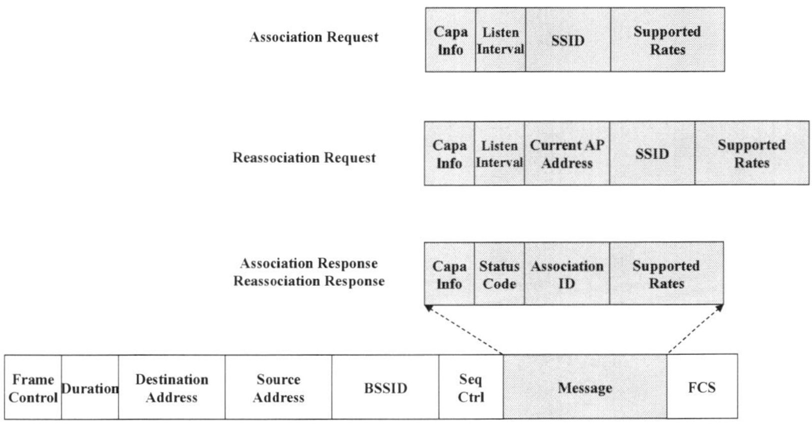

Figure 2.5. *Structure of management frames relating to the association phase*

Listen Interval: this field is coded on 16 bits. It contains the value of the number of BEACON frames during which the station will remain in standby. The access point uses this information to estimate the size of the buffer needed to store the data.

Current AP Address: this field is coded on six bytes and contains the MAC address of the access point. This field is used when the station changes the access point. It indicates the address of the old access point to the new one so that the latter can retrieve the stored data.

AID: this field is coded on 16 bits and contains the identifier of the station allocated by the access point.

The DISASSOCIATION and DEAUTHENTICATION management frames are used to terminate association and authentication, respectively. They contain a 16-bit field, indicating the reason for the shutdown (Figure 2.6).

Frame Control	Duration	Destination Address	Source Address	BSSID	Seq Ctrl	Message Reason Code	FCS

Figure 2.6. *Structure of the management frames DISASSOCIATION and DEAUTHENTICATION*

2.2. Procedures

2.2.1. *Timers*

The transmission of several frames is separated by an inter-frame interval. Several types of intervals are defined, each determining a priority level:

– SIFS (Short Inter-Frame Space): this interval corresponds to the highest priority level. It is used following the RTS and CTS control frames and the traffic frame;

– DIFS (DCF Inter-Frame Space): this interval has a longer duration (DIFS = SIFS + 2 ST (Slot Time)). It is used following an ACK control frame when the traffic frame has been correctly received;

– EIFS (Extended Inter-Frame Space): this interval is used when the transmitter has not received an acknowledgment. Its duration is equal to SIFS + (8 × ACK) + (PLCP header) + DIFS.

2.2.2. *Mobile registration*

Mobile registration at the access point is done in three phases, namely scanning, authentication and association.

The purpose of scanning is to recover the characteristics of the radio interface, which can be either passive or active.

When the scanning is passive, the station scrutinizes the BEACON channel on each radio channel.

When the scanning is active, the station transmits a PROBE REQUEST management frame, containing the network identifier (SSID), broadcast to all the access points, using a broadcast BSSID.

The access point recognizable in the SSID corresponds to the PROBE RESPONSE management frame containing the characteristics of the radio interface of the access point (Figure 2.7).

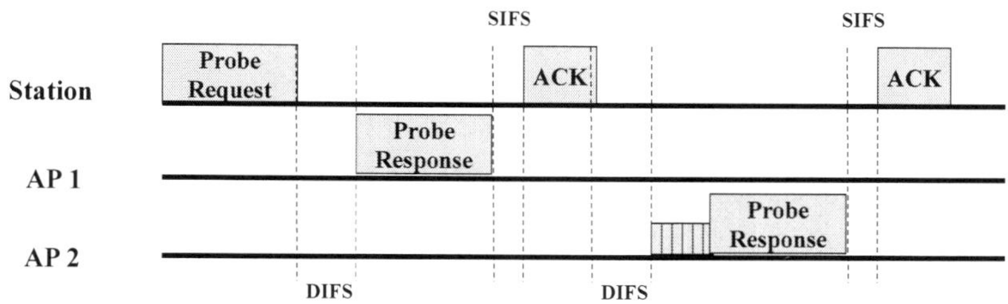

Figure 2.7. *Active scanning*

In OSA mode, authentication is done in two steps:

– the station sends the AUTHENTICATION management frame by mentioning the authentication mode;

– the access point responds with the AUTHENTICATION management frame containing the status (success or failure).

In SKA mode, authentication is done in four steps:

– the station sends the AUTHENTICATION management frame by mentioning the authentication mode;

– the access point sends the AUTHENTICATION management frame containing a bit string in the Challenge Text field;

– the station sends the AUTHENTICATION management frame containing the encrypted bit string in the Challenge Text field;

– the access point verifies the response of the station and sends the AUTHENTICATION management frame containing the status (success or failure).

The aim of the association phase is to check that the transmission characteristics of each part (the station, the access point) are compatible. It is carried out in two phases:

– the station sends the ASSOCIATION REQUEST management frame;

– the access point sends the ASSOCIATION RESPONSE management frame containing the AID assigned to the station and the status (success or failure).

The cell change is initiated by the station, by issuing the REASSOCIATION REQUEST management frame to a new access point. This frame contains the MAC address of the old access point.

The new access point responds with the REASSOCIATION RESPONSE management frame that contains the new identifier (AID) assigned to the station. In the meantime, the station must perform an authentication phase.

2.2.3. *Data transfer*

The distributed coordination function (DCF) mode implements the CSMA/CA (Carrier Sense Multiple Access/Collision Avoidance) mechanism. A station first listens to the radio channel before transmitting. To avoid collisions, the backoff mechanism is used before transmission of a frame if the radio channel is busy. The use of RTS and CTS control frames makes it possible to limit the impact of a collision to the single short RTS frame (Figure 2.8).

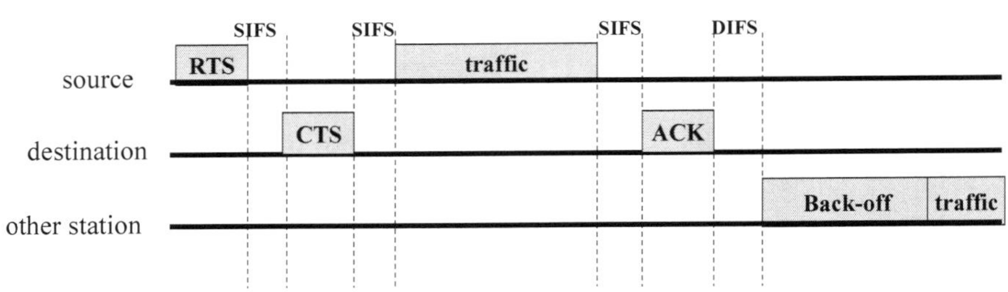

Figure 2.8. *Use of control frames for data transfer*

The use of RTS and CTS control frames, for the transmission of unicast frames from the station and the access point, and multicast or broadcast frames from the station, depends on a configuration parameter, corresponding to the size of the frame.

The multicast or broadcast frames transmitted by the access point are transmitted without RTS and CTS control frames.

The transmitted unicast traffic frames must be acknowledged by an ACK control frame (Figure 2.8), as well as multicast or broadcast traffic frames sent by the station.

The multicast or broadcast traffic frames sent by the access point are not acknowledged.

When a frame is sent, the transmitter arms a timer. If the acknowledgment is not received when the latter expires, the transmitter will try retransmitting again using the EIFS interval.

If the radio channel is available for a longer time than the DIFS, the station can transmit without the backoff timer.

If the radio channel is busy and another station wishes to transmit, it must use the backoff timer, which is the product of a random number and the time of the time slot (ST) (Figure 2.9).

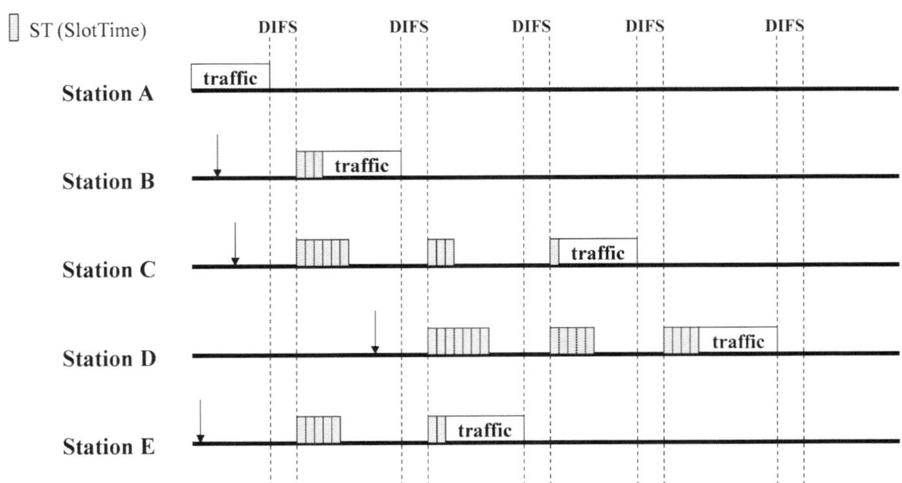

Figure 2.9. *Backoff mechanism*

This timer avoids the collision, which occurs only when two stations have drawn the same random number.

At startup, the random number is chosen in the contention window between 0 and 15. At each collision, the contention window is doubled until it reaches the maximum value 1023.

The radio channel is declared inaccessible after N access attempts, N being a parameter of the transmitter.

For a station, the consumption of this timer stops when the radio resource has been allocated. It resumes when the resource becomes free after the DIFS timer.

2.2.4. *Clear channel assessment*

Clear channel assessment (CCA) is determined at the physical level or at the logical level. At the physical level, the station is based on the detection of energy or the carrier in the radio channel. At the logical level, the station uses the Duration field of the MAC header. Logical level detection solves the problem of hidden stations.

If two stations A and B are separated by an obstacle, these two stations being connected to the same access point, each station cannot detect a transmission from the other station. The frames coming from the access point and containing the Duration field provide each station with an indication of the occupancy time of the radio channel.

The Duration field of the RTS frame contains the occupancy time of the radio channel. It is equal to the sum of the duration of three SIFS intervals, CTS and ACK control frames and a traffic frame (Figure 2.10).

The Duration field of the CTS frame contains an update of the occupancy time of the radio channel. It is equal to that indicated by the RTS frame minus the sum of the durations of one SIFS interval and CTS control frame (Figure 2.10).

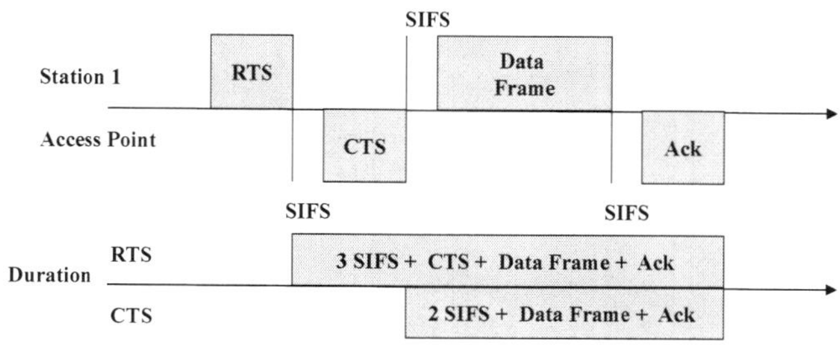

Figure 2.10. *Duration field for RTS and CTS control frames*

The Duration field of the ACK frame is set to ZERO in the case where the bit More Fragment is ZERO. In the case where this bit is at ONE, it contains the occupancy time of the radio channel for the transmission of the next fragment. It is equal to the sum of the durations of two SIFS intervals, a fragment and an ACK control frame (Figure 2.11).

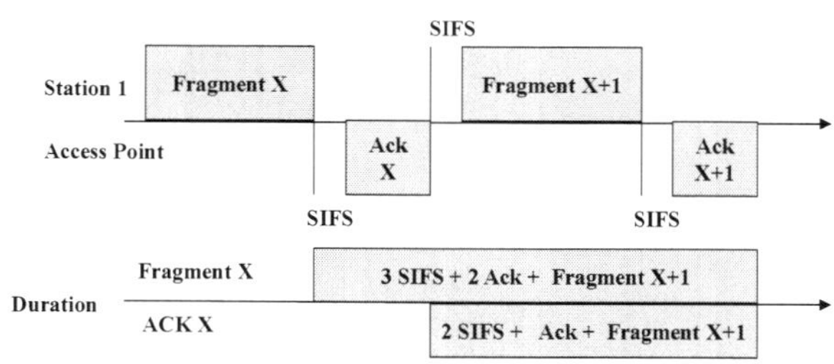

Figure 2.11. *Duration field for ACK control frame*

A station wakes up to recover data stored at the access point. It does not know the size of the pending data. The Duration field of the PS-POLL frame contains only the duration of one SIFS interval and an ACK control frame (Figure 2.12).

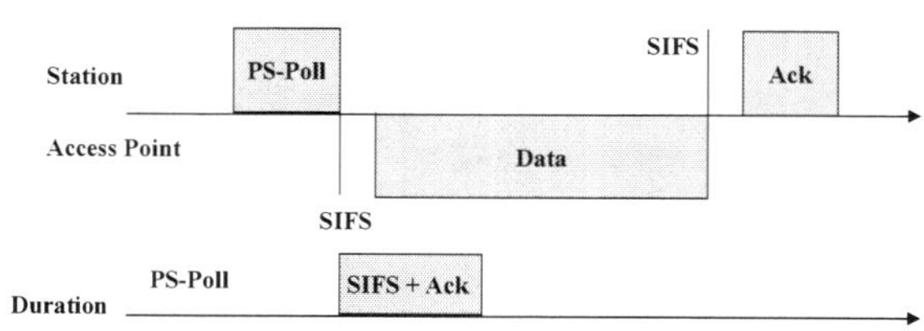

Figure 2.12. *Duration field for the PS-POLL control frame*

2.2.5. *Frame fragmentation*

A MAC frame can be fragmented to reduce the impact of interference. In the case of a faulty reception, only the corrupted fragment is retransmitted, rather than the complete frame.

Frame fragmentation is performed if the length of the frame is greater than a value defined by the network administrator. Fragments have the sequence number at the same value. The fragment number is incremented by one unit for each transmitted fragment. The More Fragment field (value of ONE) of the MAC header tells the receiver if other fragments will be transmitted.

Fragments and control frames are separated by one SIFS interval to immobilize the radio channel during transmission of all fragments (Figure 2.13).

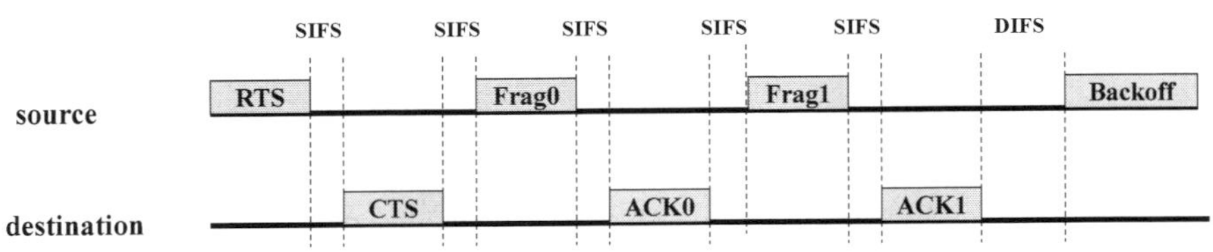

Figure 2.13. *Frame fragmentation*

2.2.6. *Standby management*

In order to save the battery consumption and increase its autonomy, the station can implement the standby mechanism. The station notifies the access point by transmitting a traffic frame (possibly a null frame) with the Power Management bit set to ONE.

The station regularly listens for a beacon frame according to the information provided in the Listen Interval field of the ASSOCIATION management frame.

The access point informs the station, via its BEACON channel, whether unicast, multicast or broadcast frames are pending (Figure 2.14). It is possible that, in the transmit interval of the beacon channel, the radio channel is busy. In this case, the station must extend the period of activity for the treatment of the beacon channel.

TIM (Traffic Indication Map): this optional field of the BEACON management frame indicates to a station in standby that data is pending. These data are either unicast data (TIM information) or multicast data (DTIM information). This field has a variable length less than or equal to 256 bytes and is composed of the following subfields:

– DTIM Count: this subfield indicates the number of BEACON frames before the next DTIM. When the value is zero, the TIM field contains DTIM information;

– DTIM Period: this subfield indicates the number of BEACON frames separating two DTIM frames;

– Bitmap Control and Partial Virtual Bitmap: these two subfields allow the identification of stations with pending data by the AID parameter.

The station sends a PS-POLL frame to warn the access point. The latter can respond immediately after one SIFS or postpone the transfer. In all cases, the station must remain awake until data transfer. If more than one frame is waiting, then the access point informs the station via the More Data field in the MAC header (Figure 2.14).

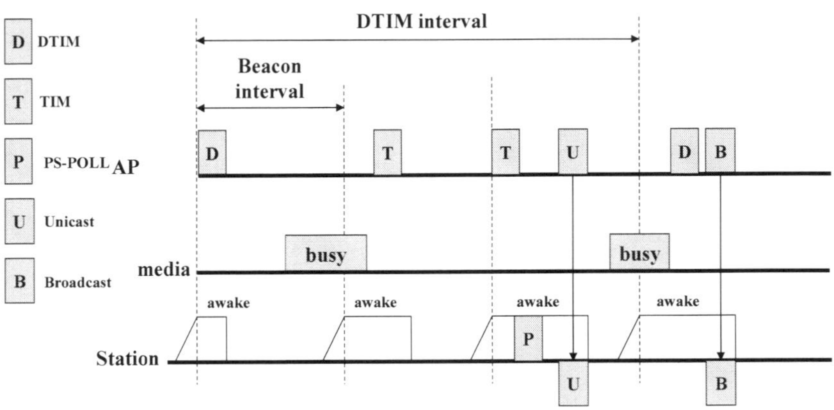

Figure 2.14. *Standby management*

2.3. Security

2.3.1. *Security mechanism*

The security of the radio interface started with the WEP mechanism. Because of its weaknesses, it was supplanted by the WPA1 mechanism and then by the WPA2 mechanism. The mechanisms WPA1 and WPA2 constitute the RSN (Robust Security Network) architecture.

The three mechanisms WEP, WPA1 and WPA2 specifically implement third-party access control and data protection services (confidentiality and integrity control).

For the WEP mechanism, third-party access control is based on the RC4 (Rivest Cipher) algorithm. Access control takes place during the authentication phase, which is a procedure associated with the MAC data link protocol.

The WPA1 and WPA2 mechanisms use the 802.1x mechanism described in Chapter 6 for access control. The authentication WPA phase is preceded by the procedure putting the security policy in agreement between the access point and the station, during the association phase.

For the WEP and WPA1 mechanisms, encryption is executed by the RC4 algorithm. For the WEP mechanism, the master key (MK) is used for the encryption of each Wi-Fi frame. For the WPA1 mechanism, encryption is obtained using a temporary key derived from the MK.

In association with encryption, a protocol is added to the MAC data link layer:

– WEP protocol in the case of the WEP mechanism;

– temporal key integrity protocol (TKIP) in the case of the WPA1 mechanism.

For the WPA2 mechanism, encryption is based on the advanced encryption standard (AES) algorithm and the header of the MAC data link protocol is completed by the CCMP (Counter-mode/Cipher block chaining MAC (Message Authentication Code) Protocol) header.

In the case of the WEP mechanism, the integrity control is provided by a cyclic redundancy check (CRC) encrypted with the RC4 algorithm.

In the case of the WPA1 mechanism, integrity control uses the MICHAEL algorithm.

In the case of the WPA2 mechanism, integrity control is obtained using the AES encryption algorithm.

2.3.2. *Security policies*

The security policies supported by the access point are transmitted in BEACON and PROBE RESPONSE frames during the scanning phase.

The response of the station to the security policies supported is included in the ASSOCIATION REQUEST frame during the association phase. This frame is validated by the ASSOCIATION RESPONSE frame from the access point.

Security policies are transmitted in RSN information element (IE).

2.3.3. *MAC header extension*

2.3.3.1. *WEP protocol*

The WEP protocol adds eight bytes to the MAC header (Figure 2.15):

– WEP header is composed of initialization vector (IV) and KeyID fields:

- IV field, coded on three bytes, is the initialization vector used to generate the pseudo-random sequence of the RC4 algorithm;

- KeyID field, coded on two bits, enables the selection of a key from among four possible ones;

– integrity check value (ICV) field, coded on four bytes, is the result of the calculation of a CRC-32 applied to MAC service data unit (MSDU). The LLC frame constitutes the MSDU data.

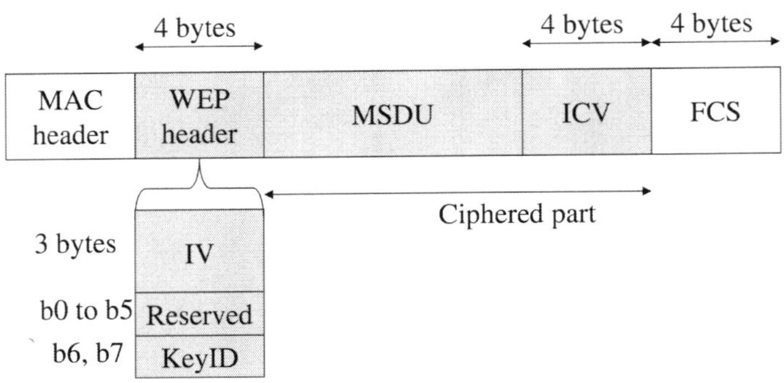

Figure 2.15. *Format of WEP encapsulation*

The 128-bit (or 64-bit) secret is composed of a 104-bit (or 40-bit) WEP key concatenated with the 24-bit IV. The secret determines the start sequence of the pseudo-random sequence of the RC4 algorithm (Figure 2.16).

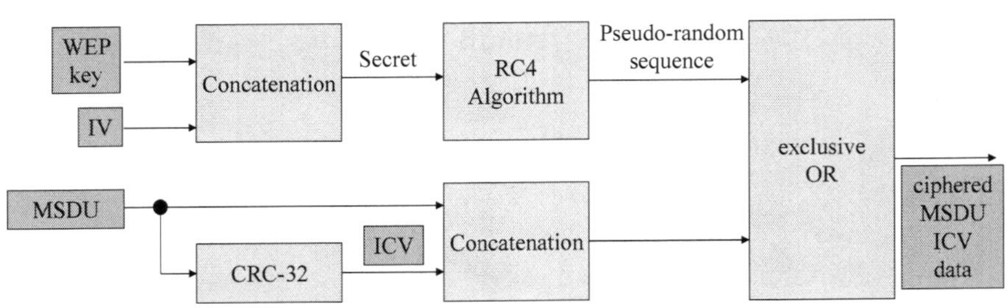

Figure 2.16. *WEP processing of the transmission chain*

Encryption consists of executing an exclusive OR of data, including the MSDU and ICV fields, on the one hand, and the pseudo-random sequence of the RC4 algorithm, on the other.

Upon reception of the MAC frame, the following operations are executed (Figure 2.17):

– the secret is reconstituted from the WEP key and the IV field;

– the pseudo-random sequence is initialized using the secret;

– the unscrambled data (MSDU and ICV) are generated by the exclusive OR of the encrypted data and the pseudo-random sequence;

– the local calculation of the CRC-32 on the MSDU data is compared to the ICV field received. If the two values are equal, then the MSDU data integrity check is positive; otherwise, the MSDU data are deleted.

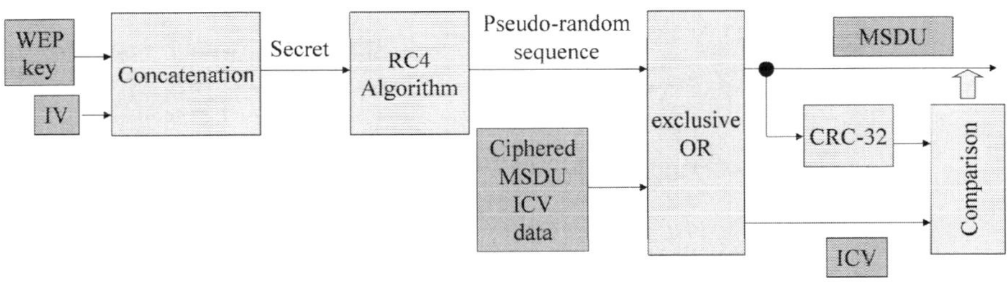

Figure 2.17. *WEP processing of the reception chain*

2.3.3.2. *TKIP*

The TKIP reuses the format of the WEP protocol. It adds four bytes to the WEP header to introduce an extension of the initialization vector. It adds eight bytes to the MSDU data to join the MIC (Message Integrity Code) field containing the seal calculated using the MICHAEL algorithm.

The TKIP header is composed of the following fields (Figure 2.18):

– the TSC0 and TSC1 (TKIP Sequence Counter) fields constitute the initialization vector and are used during the second phase of the key mixing (key derivation) function;

– the TSC2 to TSC5 fields constitute an extension of the initialization vector and are used during the first phase of the key mixing (key derivation) function;

– the WEPseed field is calculated from the TSC1 field;

– the ExtIV field, coded on one byte, indicates the presence (bit set at ONE) of the TSC2 to TSC5 fields of the IV extension;

– as for the WEP protocol, the KeyID field, coded on two bits, enables the selection of one key from among four possible ones.

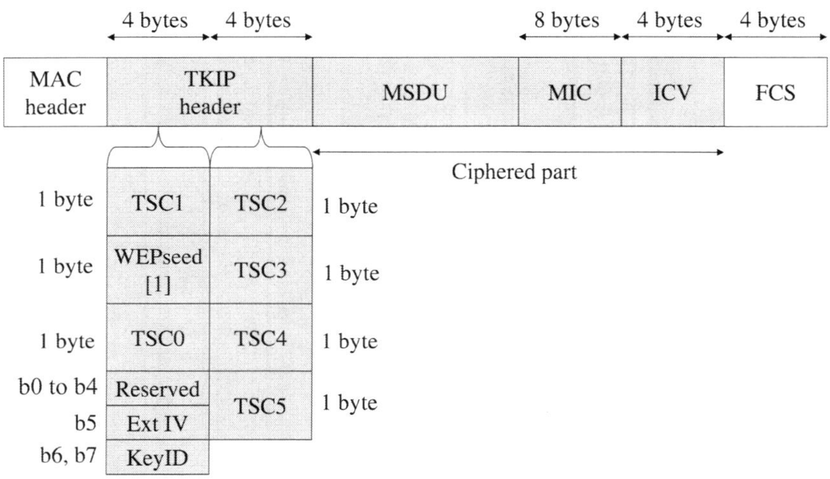

Figure 2.18. *Format of TKIP encapsulation*

The processing of the transmission chain is shown in Figure 2.19.

The MIC seal is calculated using the MAC addresses of the source and the destination, the priority byte and the MSDU data. The priority byte contains the priority level of the frame. The generation of the MIC or TMK is described in section 6.2.1.

The set composed of MSDU and MIC data can be fragmented. In this case, the initialization vector is increased by one unit for each fragment. Conversely, the IV extension keeps the same value for all fragments of a single MSDU.

For each MSDU, two key mixing (key derivation) phases are used to calculate the secret used for WEP processing:

– the first phase operates using the transmit address (TA), the TK and the TSC2 to TSC5 vectors;

– the second phase operates using the TTAK (TKIP-mixed Transmit Address and Key) key, the TK, and the TSC0 and TSC1 vectors. The TTAK constitutes an intermediary key produced during phase one.

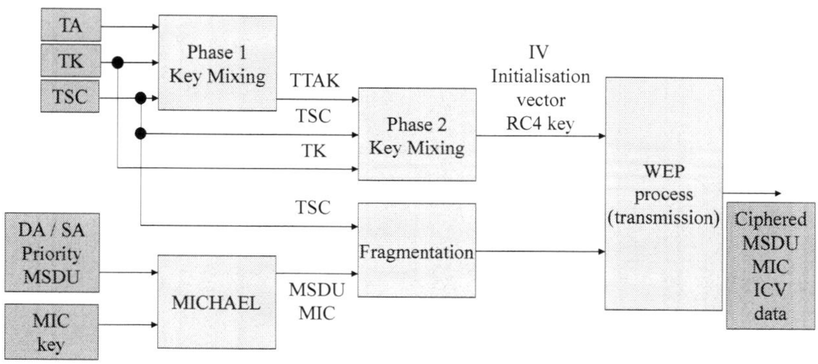

Figure 2.19. *TKIP processing of the transmission chain*

The processing of the reception chain is shown in Figure 2.20.

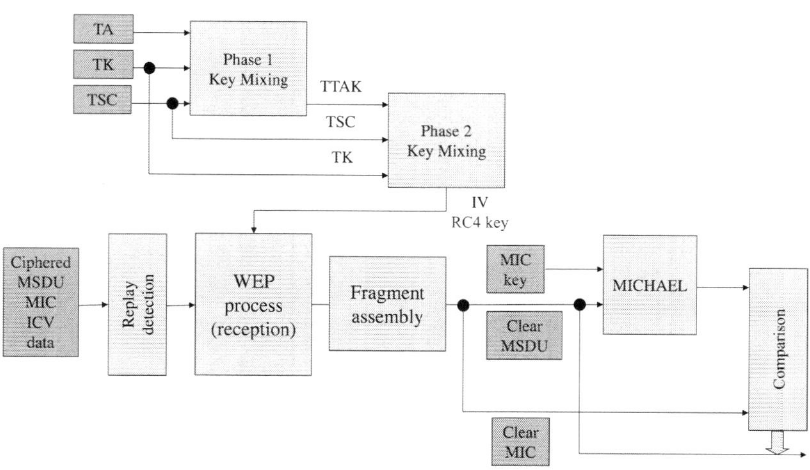

Figure 2.20. *TKIP processing of the reception chain*

The receiver extracts the TSC fields from the TKIP header and verifies the sequencing in order to protect itself from replays.

The combination of the TSC fields with the TK and the MAC transmit address (TA) enables the initialization vector and the RC4 key to be reconstituted for the WEP decryption.

If the WEP processing indicates a positive check from the ICV field, the fragments are reassembled.

The result of the MIC seal calculation using the MIC key and unscrambled MSDU data is compared to the value of the MIC field received. If the two values match, the integrity control is positive and the MSDU data are accepted.

2.3.3.3. *CCMP*

The CCMP adds 16 bytes to the MAC header (Figure 2.21):

– eight bytes for the CCMP header;

– eight bytes for the MIC seal.

The CCMP header resembles the TKIP header. It is constructed from the packet number (PN), the ExtIV field, coded on one bit and indicating the presence (bit set at ONE) of the PN2 to PN5 fields, and the KeyID field, coded on two bits, used to select a key from among four possible keys.

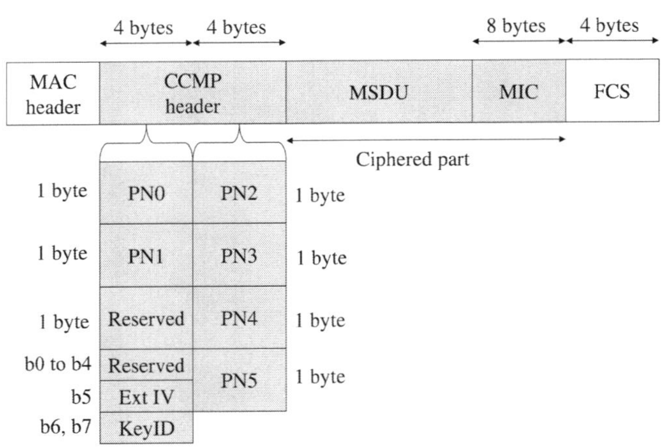

Figure 2.21. *Format of CCMP encapsulation*

The processing of the transmission chain is shown in Figure 2.22.

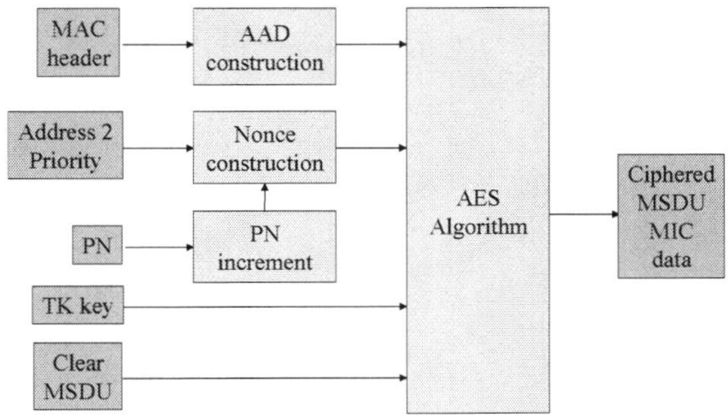

Figure 2.22. *CCMP processing of the transmission chain*

The AES algorithm provides the MIC seal and encryption of the MSDU and MIC data. It is supplied by the following values:

– AAD (Additional Authentication Data) parameter, built from the MAC header, with the exception of fields that can be modified during a retransmission (e.g. the Duration field);

– Nonce parameter, built from the priority byte, the second address contained in the MAC header (A2) and the frame number (PN). The value of the PN field is increased by one unit for each frame generated;

– TK.

The processing of the CCMP received chain is shown in Figure 2.23.

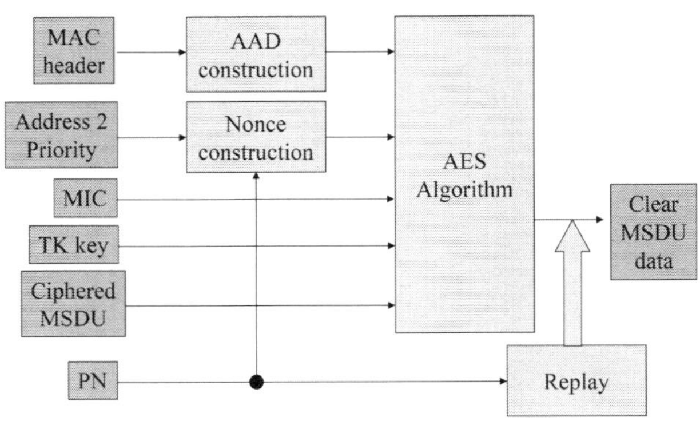

Figure 2.23. *CCMP processing of the reception chain*

The AES algorithm is used to reproduce unscrambled MSDU data. It is supplied by the following values:

– AAD parameter;

– Nonce parameter;

– MIC field to execute an integrity control;

– TK.

A check on the PN field enables protection from replays.

2.4. Quality of service

2.4.1. *EDCA mechanism*

The EDCA (Enhanced Distributed Channel Access) mechanism is an extension of the DCF mechanism. It is based on the differentiation of user priority (UP) levels. The priority level is the same as that set for the Ethernet frame.

The ability to implement the EDCA mechanism is indicated by the access point in the Capability Information field of the BEACON or PROBE RESPONSE management frames.

The EDCA mechanism defines four access categories (AC), each category corresponding to a queue (Table 2.3):

– a priority level belongs to an access category;

– an access category contains two priority levels.

TID field	User Priority UP	Access Category AC	Designation
1	BK	AC-BK	Background
2	–	AC-BK	Background
0	BE	AC-BE	Best Effort
3	EE	AC-BE	Best Effort
4	CL	AC-VI	Video
5	VI	AC-VI	Video
6	VO	AC-VO	Voice
7	NC	AC-VO	Voice

Table 2.3. *Correspondence between the priority levels and the access categories*

The EDCA mechanism defines the parameters used for access to the radio channel:

– the arbitration inter-frame space (AIFS) during which the mobile detects that the radio channel is free, before triggering the backoff mechanism or the transmission:

AIFS[AC] = AIFSN[AC] × SlotTime + SIFSTime;

– the minimum length of the contention window CWmin (Contention Window) and the maximum CWmax used for the backoff mechanism;

– TXOP (Transmission Opportunity) time during which the mobile transmits when it has access to the radio channel.

The EDCA parameters are stored in the mobile and can be updated by the access point in the EDCA Parameter Set information element transmitted in the BEACON, ASSOCIATION RESPONSE, REASSOCIATION RESPONSE and PROBE RESPONSE management frames.

The modification of the EDCA parameters is indicated in the QoS Capability information element present in the BEACON management frame that does not contain the EDCA Parameter Set information element and in the (RE)ASSOCIATION REQUEST management frames (RE).

The default values of the EDCA parameters are shown in Table 2.4.

Access category	CWmin	CWmax	AIFSN	TXOP (ms)
AC-BK	15	1,023	7	0
AC-BE	15	1,023	3	0
AC-VI	7	15	2	3,008
AC-VO	3	7	2	1,504

Table 2.4. *Default values of EDCA parameters*

2.4.2. *Impact on the MAC header*

The MAC header of the traffic frame inserts the two-byte QoS Control field following the Address 4 field (Figure 2.24).

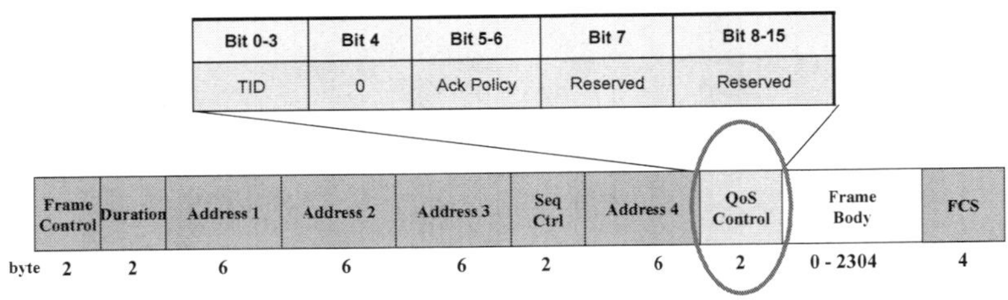

Figure 2.24. *Evolution of MAC header structure*

The presence of the QoS Control field is indicated by the Subtype field (value equal to 1,000).

The TID (Traffic Identifier) subfield, coded on three bits, provides the priority level (UP) of the access to the radio channel.

The Ack Policy subfield, coded on two bits, identifies the MAC frame acknowledgment rule:

– 00: the frame must be acknowledged. The receiver of the MAC frame must return an ACK frame after the SIFS interval;

– 01: the frame must not be acknowledged. Acknowledgment is performed in the upper layers. This combination is also used for multicast or broadcast frames;

– 11: the block acknowledgment mechanism must be used.

3

802.11a/g Interfaces

3.1. 802.11a interface

3.1.1. *PLCP sub-layer*

On transmission, the physical layer convergence procedure (PLCP) converts the PLCP service data units (PSDU) from the MAC (Medium Access Control) layer to form the PLCP protocol data units (PPDU), adding a preamble and a header.

On reception, the preamble and header facilitates demodulation of the signal and the delivery of the PSDU data units.

The PLCP frame ends with tail and padding bits. The PLCP header contains the LENGTH and RATE fields, a reserved bit, an even parity bit and the SERVICE field (Figure 3.1).

The LENGTH and RATE fields, the reserved bit and the even parity bit, the SIGNAL set, constitute an OFDM (Orthogonal Frequency-Division Multiplexing) symbol and are transmitted with the most robust modulation and coding scheme:

– BPSK (Binary Phase-Shift Keying) modulation;

– coding rate of 1/2.

The SERVICE field, the PSDU data units, the tail and padding bits, the DATA set, are transmitted at the rate indicated in the RATE field and constitute several OFDM symbols (Table 3.1).

The first seven bits of the SERVICE field are used to synchronize the descrambler on reception. The other nine bits are reserved for later use.

The LENGTH field encodes the number of bytes of the PSDU data unit.

The tail bits of the SIGNAL symbol allow the decoding of the RATE and LENGTH fields. The RATE and LENGTH fields allow the mobile to predict the duration of the PLCP frame, even if the bit rate is not supported by the mobile.

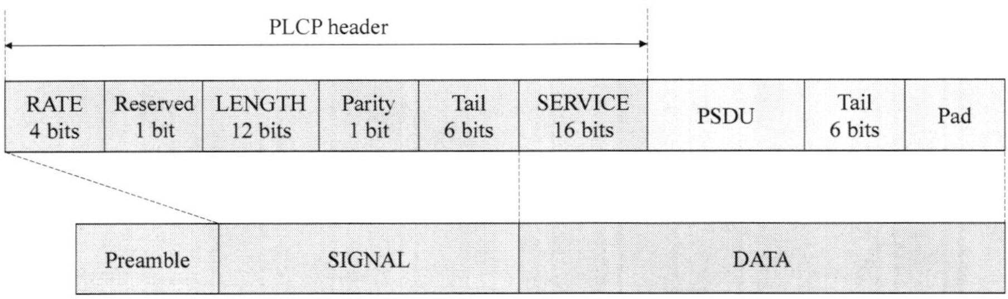

Figure 3.1. *Format of PLCP frame*

RATE field	Rate (Mbps)
1101	6
1111	9
0101	12
0111	18
1001	24
1011	36
0001	48
0011	54

Table 3.1. *Rates of DATA field*

3.1.2. *PMD sub-layer*

3.1.2.1. *Transmission chain*

The transmission chain consists of the following operations (Figure 3.2):

a) Produce the PLCP preamble field, composed of:

– ten repetitions of a short training sequence, used for ACG (Automatic Control Gain) convergence, diversity selection, timing acquisition and coarse frequency acquisition in the receiver;

– two repetitions of a long training sequence, used for channel estimation and fine frequency acquisition in the receiver, preceded by a guard interval (GI).

b) Produce the PLCP header field from the RATE, LENGTH and SERVICE fields by filling the appropriate bit fields.

c) Calculate from the RATE field the number of data bits per OFDM symbol (N_{DBPS}), the coding rate (R), the number of bits in each OFDM sub-carrier (N_{BPSC}) and the number of coded bits per OFDM symbol (N_{CBPS}).

d) Append the PSDU data unit to the SERVICE field. Extend the resulting bit string with bits to ZERO (at least six bits) so that the resulting length is a multiple of N_{DBPS}. The resulting bit string constitutes the DATA part of the frame.

e) Initiate the scrambler with a pseudo-random non-zero seed, generate a scrambling sequence and XOR it with the extended string of data bits.

f) Replace the six scrambled bits at ZERO following the data with six non-scrambled bits at ZERO. Those bits return the convolutional encoder to the zero state and are denoted as tail bits.

g) Encode the extended, scrambled data string with a convolutional encoder (R = 1/2). Omit some of the encoder output string, chosen according to puncturing pattern, to reach the desired coding rate.

h) Divide the encoded bit string into groups of N_{CBPS} bits. Within each group, perform an interleaving of the bits according to a rule corresponding to the desired RATE.

i) Divide the resulting coded and interleaved data string into groups of N_{BPSC} bits. For each of the bit groups, convert the bit group into a complex number according to the modulation encoding tables.

j) Divide the complex number string into groups of 48 complex numbers. Each such group is associated with one OFDM symbol. In each group, the complex numbers are numbered 0 to 47 and mapped hereafter onto OFDM sub-carriers numbered −26 to −22, −20 to −8, −6 to −1, 1 to 6, 8 to 20 and 22 to 26. The sub-carriers −21, −7, 7 and 21 are skipped and, subsequently, used for inserting pilot sub-carriers. The 0 sub-carrier, associated with center frequency, is omitted and filled with the value 0. The total number of the sub-carriers is 52 (48 + 4).

k) For each group of sub-carriers −26 to 26, convert the sub-carriers to the time domain using inverse Fourier transform. Prepend to the Fourier-transformed waveform a circular extension of itself, thus forming a GI, and truncate the resulting periodic waveform to a single OFDM symbol length by applying time domain windowing.

l) Append the OFDM symbols one after another, starting after the SIGNAL symbol describing the RATE and LENGTH fields.

m) Up-convert the resulting complex baseband waveform to a radio signal according to the center frequency of the desired channel and transmit.

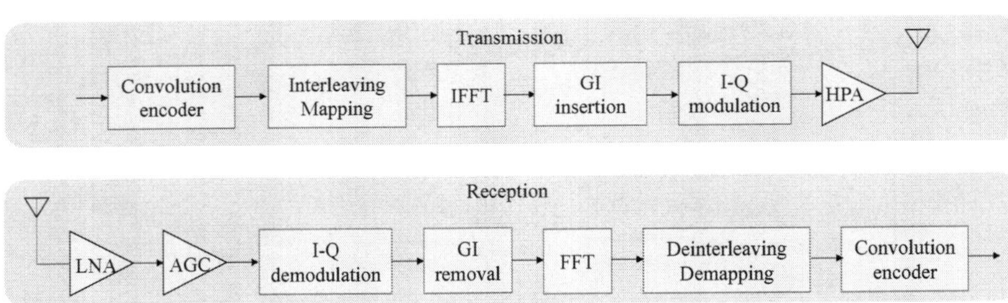

Figure 3.2. *Transmission and reception chain*

3.1.2.2. *Scrambler*

The DATA field, composed of SERVICE, PSDU data unit, tail and pad parts, shall be scrambled with a length-127 scrambler (Figure 3.3).

The same scrambler is used to scramble transmit data and to descramble receive data. When transmitting, the initial state of the scrambler shall be set to a pseudo-random non-zero state.

The seven least significant bits (LSB) of the SERVICE field shall be set to all zeros prior to scrambling to enable estimation of the initial state of the scrambler in the receiver.

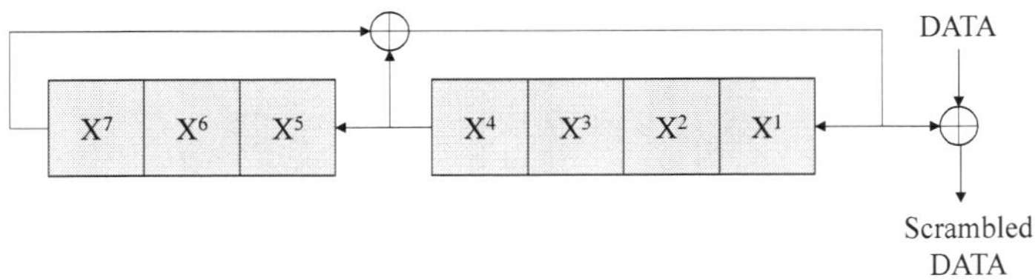

Figure 3.3. *Scrambler diagram*

3.1.2.3. *Convolutional encoder*

The convolutional encoder shall use the generator polynomials $g0 = 133_8$ and $g1 = 171_8$ and produce two sequences from the scrambled DATA field (Figure 3.4).

The rates are derived from the two sequences by employing puncturing. Puncturing is a procedure for omitting some of the encoded bits in the transmitter and inserting bits at ZERO into the convolutional decoder on the receive side in place of the omitted bits.

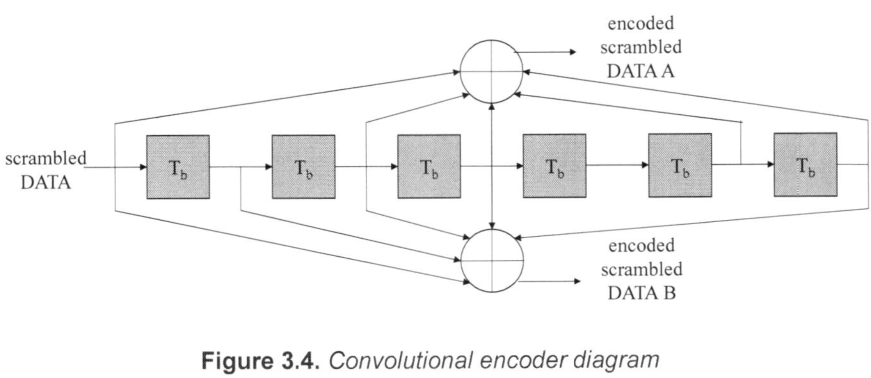

Figure 3.4. *Convolutional encoder diagram*

3.1.2.4. *Interleaving*

All encoded data bits shall be interleaved by a block interleaver with a block size corresponding to the number of bits in a single OFDM symbol (N_{CBPS}).

The interleaver is defined by a two-step permutation. The first permutation ensures that adjacent coded bits are mapped onto non-adjacent sub-carriers. The second ensures that adjacent coded bits are mapped alternately onto less and more significant bits of the constellation, and therefore long runs of low-reliability (LSB) bits are avoided.

3.1.2.5. *Modulation and coding scheme*

The modulation and coding scheme determines the rate of the DATA field (Table 3.2).

The sub-carriers are modulated using phase modulation (BPSK or QPSK) or mixed phase and amplitude modulation (16-QAM or 64-QAM).

Forward error correction (FEC) is a convolution code used with a coding rate of 1/2, 2/3 or 3/4.

Modulation	Coding rate	Number of bits per sub-carrier	Number of DATA bits per OFDM symbol	Number of encoded DATA bits	Rate (Mbps)
BPSK	1/2	1	24	48	6
BPSK	3/4	1	36	48	9
QPSK	1/2	2	48	96	12
QPSK	3/4	2	72	96	18
16-QAM	1/2	4	96	192	24
16-QAM	3/4	4	144	192	36
64-QAM	2/3	6	192	288	48
64-QAM	3/4	6	216	288	54

Table 3.2. *Parameters of the modulation and coding scheme*

3.1.2.6. *Structure of the preamble and OFDM symbols*

The preamble consists of a short learning sequence of T_{SHORT} duration and a long learning sequence of T_{LONG} duration. The short learning sequence contains 10 short symbols t_1 to t_{10}. The long learning sequence consists of a guard time T_{GI2} and two long symbols T_1 and T_2 (Figure 3.5 and Table 3.3).

The SIGNAL symbol, of T_{SIGNAL} duration, and the different symbols DATA, of T_{SYM} duration, start with a guard time T_{GI} (Figure 3.5 and Table 3.3).

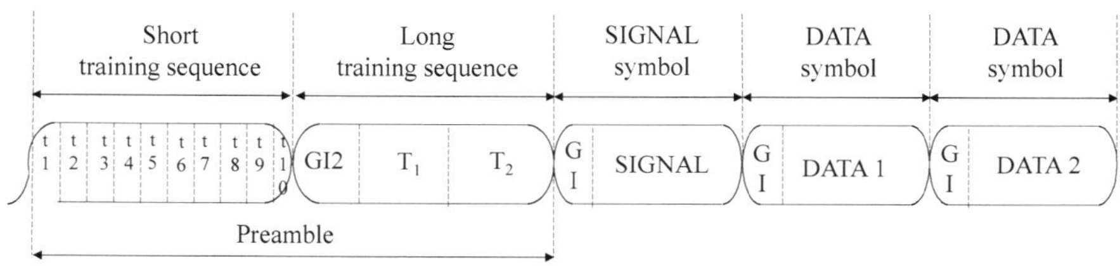

Figure 3.5. *Structure of the preamble and OFDM symbols*

Parameters	Duration
T_{FFT} FFT or IFFT period	3.2 µs ($1/\Delta_F$)
T_{GI} Duration of guard interval GI	0.8 µs ($T_{FFT}/4$)
T_{GI2} Duration of guard interval GI2	1.6 µs ($T_{FFT}/2$)
T_{SHORT} Duration of short training symbol	8 µs ($10\times T_{FFT}/4$)
T_{LONG} Duration of long training symbol	8 µs ($T_{GI2}+2\times T_{FFT}$)
$T_{PREAMBLE}$ Preamble duration	16 µs ($T_{SHORT}+T_{LONG}$)
T_{SIGNAL} Duration of SIGNAL symbol	4 µs ($T_{GI+}T_{FFT}$)
T_{SYM} Duration of DATA symbol	4 µs ($T_{GI+}T_{FFT}$)

Table 3.3. *Values of the duration of the different parameters*

A short training symbol consists of 12 sub-carriers, which are modulated by the elements of the sequence S:

$$S_{-26,26} = \sqrt{(13/6)} \times \{0, 0, 1+j, 0, 0, 0, -1-j, 0, 0, 0, 1+j, 0, 0, 0, -1-j, 0, 0, 0, -1-j, 0, 0, 0, 1+j, 0, 0, 0, 0,0, 0, 0, -1-j, 0, 0, 0, -1-j, 0, 0, 0, 1+j, 0, 0, 0, 1+j, 0, 0, 0, 1+j, 0, 0, 0, 1+j, 0,0\}$$

A long training symbol consists of 53 sub-carriers, including the value 0 at central frequency, which are modulated by the elements of the sequence L:

$$L_{-26,26} = \{1, 1, -1, -1, 1, 1, -1, 1, -1, 1, 1, 1, 1, 1, 1, -1, -1, 1, 1, -1, 1, -1, 1, 1, 1, 1, 0, 1, -1, -1, 1, 1, -1, 1, -1, 1, -1, -1, -1, -1, -1, 1, 1, -1, -1, 1, -1, 1, -1, 1, 1, 1, 1\}$$

3.1.2.7. *OFDM multiplexing*

The complex numbers are numbered from 0 to 47 and are subsequently mapped onto OFDM sub-carriers numbered −26 to −22, −20 to −8, −6 to −1, 1 to 6, 8 to 20 and 22 to 26.

Sub-carriers −21, −7, 7 and 21 are ignored and subsequently used for the insertion of pilot sub-carriers.

The sub-carrier 0, associated with the central frequency, is omitted and filled with the value 0.

Parameters	**Values**
N_{SD} (number of sub-carriers assigned to the DATA field)	48
N_{SP} (number of sub-carriers assigned to pilots)	4
N_{ST} (total number of sub-carriers)	52
Δ_F (spacing between sub-carriers)	0.3125 MHz (20 MHz/64)

Table 3.4. *Parameters of OFDM multiplexing*

3.1.2.8. *Frequency plan*

The 802.11a interface operates in the 5 GHz U-NII (Unlicensed-National Information Infrastructure) band, divided into three sub-bands:

– sub-band A covering the frequency band 5.150–5.350 GHz;

– sub-band B covering the frequency band 5.470–5.725 GHz;

– sub-band C covering the frequency band 5.725–5.825 GHz.

The bandwidth of the radio channel is equal to 20 MHz. Table 3.5 provides, for each sub-band, the channel number and the value of the center frequency.

Sub-band A		Sub-band B		Sub-band C	
Channel number	Central frequency (GHz)	Channel number	Central frequency (GHz)	Channel number	Central frequency (GHz)
36	5.180	100	5.500	149	5.745
40	5.200	104	5.520	153	5.765
44	5.220	108	5.540	157	5.785
48	5.240	112	5.560	161	5.805
52	5.260	116	5.580		
56	5.280	120	5.600		
60	5.300	124	5.620		
64	5.320	128	5.640		
		132	5.660		
		136	5.680		
		140	5.700		

Table 3.5. *U-NII band at 5 GHz*

Countries apply their own regulations to authorized channels, authorized users and maximum power levels in these frequency ranges.

Table 3.6 summarizes the rules applied in Europe for sub-bands A (5.150–5.350 GHz) and B (5.470–5.725 GHz).

Dynamic frequency selection (DFS) allows access points to automatically select frequency channels with low levels of interference.

Transmit power control (TPC) automatically reduces output power when other networks are in range. Reduced power means reduced interference problems and increased battery capacity.

Radio channels	Maximum power	TPC function	DFS function
36 to 48	200 mW	No	No
48 to 64	200 mW	Yes	Yes
	100 mW	No	
100 to 140	1 W	Yes	Yes
	500 mW	No	

Table 3.6. *European regulations*

3.2. 802.11g interface

The 802.11g interface, called ERP (Extended Rate Physical), must be compatible with the 802.11 interface, called DSSS (Direct Sequence Spread Spectrum), which provides the bit rates at 1 and 2 Mbps.

The 802.11g interface must also be compatible with the 802.11b interface, called HR (High Rate)/DSSS, which provides the bit rates at 1, 2, 5.5 and 11 Mbps.

The 802.11g interface implements the functions of the 802.11a interface, except that it uses the ISM (Industrial, Scientific and Medical) frequency band at 2.4 GHz.

The ERP access point is able to work in any combination of ERP and non-ERP modes. For example, the access point could operate only in an ERP-OFDM mode, in a mixed mode ERP-OFDM and ERP-HR/DSSS or in a mixed mode ERP-HR/DSSS/and not ERP.

The DSSS/OFDM mode uses an 802.11b-compatible header and a payload using OFDM multiplexing.

3.2.1. *PLCP sub-layer*

An ERP station must support three different formats for the PLCP header.

The first format corresponds to the ERP-HR/DSSS mode. It includes the long preamble defined for the 802.11b interface, with a redefinition of the reserved bits.

The second format corresponds to the ERP-HR/DSSS mode. It includes the optional short preamble for the 802.11b interface.

The third format corresponds to the ERP-OFDM mode. It includes the preamble and header defined for the 802.11a interface.

The DSSS-OFDM mode uses a format that includes the short or long preamble and the header of the 802.11b interface associated with the preamble and the header of the 802.11a interface. This mode is optional.

3.2.1.1. *ERP-HR/DSSS mode*

The structure of the PLCP frame is described in Figure 3.6.

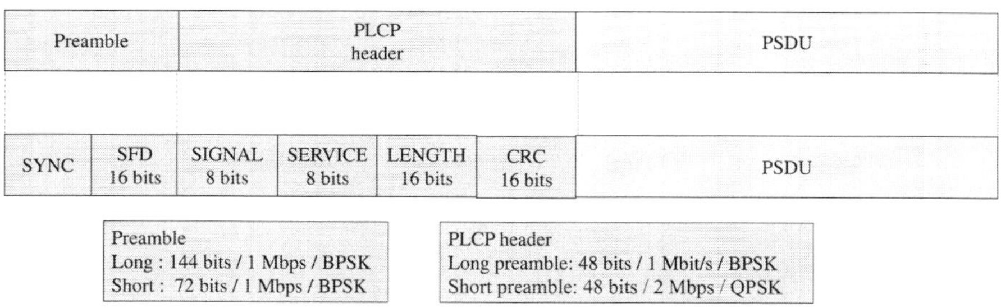

Figure 3.6. *PLCP frame for ERP-HR / DSSS mode*

The difference with the PLCP header of the 802.11b interface is in the SERVICE field bits set to support the optional packet binary convolutional code (PBCC):

– bits b0, b1 and b4 are reserved and must be set to ZERO;

– bit b2 is used to indicate that the transmission frequency and the symbol clock are derived from the same oscillator. For ERP mode, the locked clock bit must be set to ONE;

– bit b3 is used to indicate whether the data uses the PBCC option;

– bits b5, b6 and b7 are used to resolve DATA field length ambiguities for the optional PBCC mode;

– bit b7 is also used to resolve DATA field length ambiguities for CCK (Complementary Code Keying) mode;

– bits b3, b5 and b6 are set to ZERO for the CCK mode.

3.2.1.2. *ERP-OFDM mode*

The structure of the PLCP frame is described in Figure 3.7.

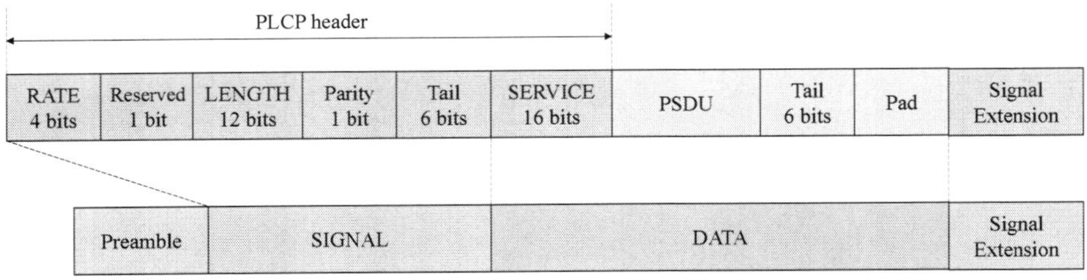

Figure 3.7. *PLCP frame for ERP-OFDM mode*

The PLCP frame is followed by a period (Signal Extension) without transmission with a duration of 6 μs.

The SIFS time for the 802.11a interface is equal to 16 μs to allow additional time for the convolutional decoding process to complete.

To be compatible with the 802.11b interface, the SIFS time for the 802.11g interface is equal to 10 μs. This extra length extension of 6 μs thus makes it possible to ensure that the convolutional decoding process ends.

3.2.1.3. *DSSS-OFDM mode*

The structure of the PLCP frame is described in Figure 3.8.

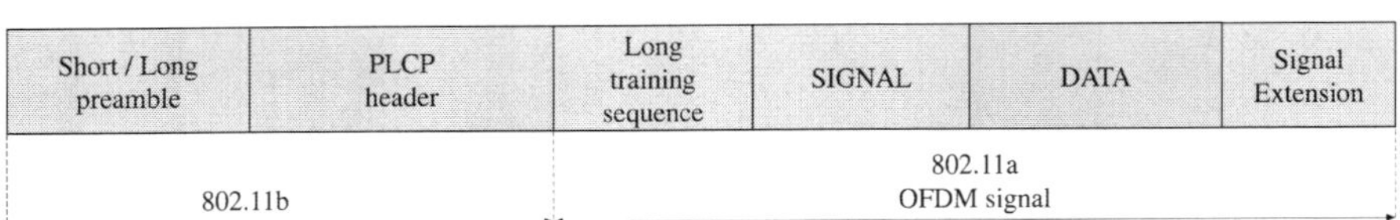

Figure 3.8. *PLCP frame for DSSS-OFDM mode*

The SIGNAL field of the PLCP header must be set to a value of 3 Mbps.

The PLCP header is similar to that described for the ERP-HR/DSSS mode.

The payload of the PLCP frame consists of a long learning sequence, the OFDM SIGNAL field, which provides information on the rate and length of the DATA field, and a signal extension section to provide additional processing time for convolutional decoding.

3.2.2. *PMD sub-layer*

The 802.11g interface operates in the ISM band at 2.4 GHz, covering the 2.4–2.4835 GHz frequency band.

The bandwidth of the radio channel is equal to 22 MHz for the 802.11b interface and 20 MHz for the 802.11g interface. Figure 3.9 shows the channel number and the value of the center frequency.

To avoid overlapping channels, it is recommended to use channels 1, 6 and 11.

Channel 14 has been designated for specific use in Japan.

The radio spectrum from 2.400 to 2.450 GHz (channels 1 to 8) is shared with radio amateurs.

Channels 1, 5, 9 and 13 are used by domestic image transmitters and analog and digital Webcams.

The 2.450 GHz frequency is that of microwave ovens that can disrupt, more or less, Wi-Fi channels 7 to 10.

The maximum authorized power, inside and outside buildings, is 100 mW.

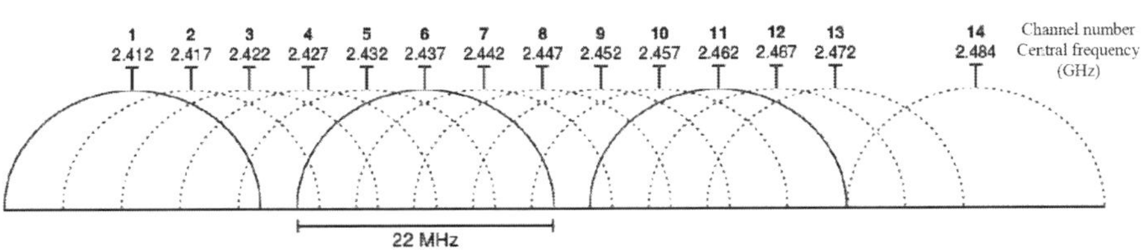

Figure 3.9. *ISM band at 2.4 GHz*
Source: http://en.wikipedia.org/wiki/IEEE_802.11g-2003

4

802.11n Interface

4.1. MAC layer evolution

Table 4.1 summarizes the features provided by the MAC (Medium Access Control) layer.

Features	Mandatory/optional	Description
Reception A-MPDU	Mandatory	MAC frame aggregation
Transmission A-MPDU	Optional	
Reception A-MSDU	Mandatory	Aggregation of MAC frame payload
Transmission A-MSDU	Optional	
Block Ack	Mandatory	Acknowledgment for a block of MAC frames
Protection	Mandatory	Detection of radio channel occupancy time by non-802.11n compatible stations
RIFS	Mandatory	Reduced inter-frame interval
Spatial Multiplexing Power Save	Mandatory	Power save by reducing the number of spatial flows
Power Save Multi-Poll	Optional	Power save by modifying the radio access procedure for smaller frames
Non–TKIP	Mandatory	TKIP is no longer allowed
Phased Coexistence Operation	Optional	Alternating radio channels at 20 and 40 MHz

Table 4.1. *Features of MAC layer*

4.1.1. *Management frames*

4.1.1.1. *HT Capabilities information element*

The management frames indicate that the access point has an 802.11n interface by including the HT (High Throughput) Capabilities information element.

The information provided by the HT Capabilities Info field is described in Table 4.2.

Information	Designation
LDPC Coding Capability	LDPC error correction code
Supported Channel Width Set	Bandwidth of the radio channel (20 MHz / 40 MHz)
SM Power Save	Power save for spatial multiplexing
HT_Greenfield	HT_GF format for PLCP header
Short GI for 20 MHz	Short guard interval for the 20 MHz radio channel
Short GI for 40 MHz	Short guard interval for the 40 MHz radio channel
Tx STBC	Transmission for the space-time diversity STBC
Rx STBC	Reception for the space-time diversity STBC
HT-Delayed Block Ack	Delayed acknowledgment mechanism
Maximum A-MSDU Length	Maximum size of frame aggregation A-MPDU (3,839 or 7,935 bytes)
DSSS/CCK Mode in 40 MHz	Using the DSSS/CCK mode for 40 MHz radio channel
Forty MHz Intolerant	Prohibition to use 40 MHz radio channel
L-SIG TXOP Protection Support	L-SIG TXOP protection mechanism

Table 4.2. *Information of HT Capabilities Info field*

A-MPDU Parameters: this field indicates the maximum size of the frame aggregation A-MPDU that the access point can receive and the minimum time between two MPDU data units of the aggregation.

Supported MCS Set: this field indicates the modulation and coding schemes supported by the access point, for transmission and reception.

HT Extended Capabilities: this field indicates whether PCO (Phased Coexistence Operation) mode or RD (Reverse Direction) protocol is supported.

Transmit Beamforming Capabilities: this field describes the supported features for beamforming.

ASEL Capability: this field describes the supported features for antenna selection.

The HT Capabilities information element is included in BEACON frames so that mobiles can determine that the 802.11n interface is available.

A mobile inserts the HT Capabilities information element into the PROBE REQUEST frame to search for 802.11n access points.

The HT Capabilities information element is also included in the management frames ASSOCIATION REQUEST, ASSOCIATION RESPONSE, REASSOCIATION REQUEST, REASSOCIATION RESPONSE and PROBE RESPONSE.

4.1.1.2. *HT Operation information element*

The HT Operation information element provides the mobile with the characteristics of the 802.11n interface and contains the following fields:

Primary Channel: this field indicates the number of the primary radio channel. This channel is used for management frames.

Secondary Channel Offset: this field indicates whether the secondary radio channel has a frequency higher or lower than that of the primary channel.

STA Channel Width: this field indicates the bandwidth that the access point uses in reception.

RIFS mode: this field indicates whether the use of the reduced inter-frame space (RIFS) is allowed.

HT Protection: this field indicates the protection mechanism to avoid interference with mobiles that are not compatible with the 802.11n interface.

Non-greenfield STA present: this field indicates if the HT_GF mode is supported by the access point.

OBSS Non-HT STAs present: this field indicates that an overlapping basic service set (OBSS) contains mobiles that are not compatible with the 802.11n interface requiring protection.

Dual Beacon, Dual CTS, STBC Beacon: these three modes are used when the beacon channel uses the diversity in STBC transmission.

L-SIG Protection Full Support: this field indicates whether the L-SIG protection mechanism is supported.

PCO Active, PCO Phase: these two fields indicate the use of the PCO mode, which makes it possible to switch a radio channel between 20 and 40 MHz. These fields are used to indicate that the PCO mode is in operation and whether the radio channel is currently 20 or 40 MHz.

Basic MCS set: this field indicates the modulation and coding schemes supported by the access point.

The HT Operation information element is included in the management frames BEACON, ASSOCIATION RESPONSE, REASSOCIATION RESPONSE and PROBE RESPONSE.

4.1.2. *Structure of the MAC header*

The 802.11n interface modifies the structure of the protocol header by adding the HT Control field (Figure 4.1) after the QoS Control field.

The presence of the HT Control field is indicated by the Order bit of the Frame Control field set to ONE, for QoS Data traffic frames and management frames.

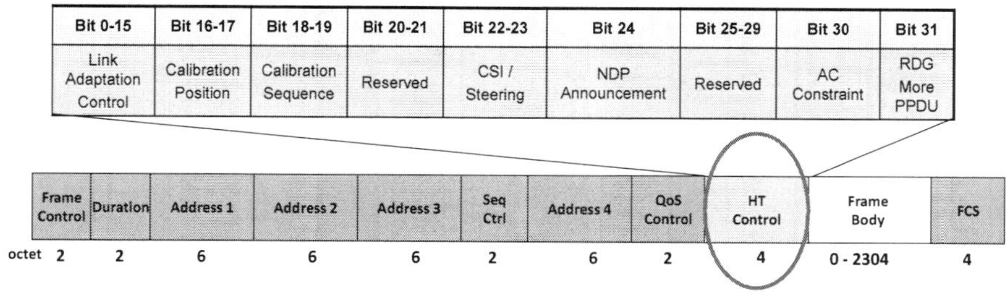

Figure 4.1. *Structure of MAC header*

The information provided by the Link Adaptation Control field is described in Table 4.3.

Information	Designation
TRQ (Training Request)	Request for the transmission of a PPDU sounding
MAI (MCS request and ASEL Indication)	Interpretation of MFB/ASELC information
MFSI (MCS Feedback Sequence Identifier)	Identifier of the sequence relating to a request for a recommendation on the value of the modulation and coding scheme (MCS)
MFB/ASELC (MCS Feedback and Antenna Selection Command)	MCS recommended value or features of antenna selection

Table 4.3. *Information of Link Adaptation Control field*

Calibration Position: this field indicates the position in the exchange sequence relative to the calibration sounding.

Calibration Sequence: this field contains the identifier of the exchange sequence.

CSI/Steering: this field indicates the type of response for beamforming.

NDP Announcement: this field indicates whether an empty frame is transmitted after the data unit.

AC Constraint: this field indicates whether the data in the RD protocol belongs to a single access category (AC).

RDG *More* PPDU: this field is interpreted differently, for the RD protocol, if it is transmitted by the initiator (allocation of a resource or not) or the responder (the frame is the last transmitted or not).

4.1.3. *Frame aggregation*

4.1.3.1. *A-MPDU frame*

The A-MPDU (Aggregate MAC Protocol Data Unit) frame is an A-MPDU sub-frame sequence (Figure 4.2). Each A-MPDU sub-frame contains a delimiter, an MPDU frame and pad bytes.

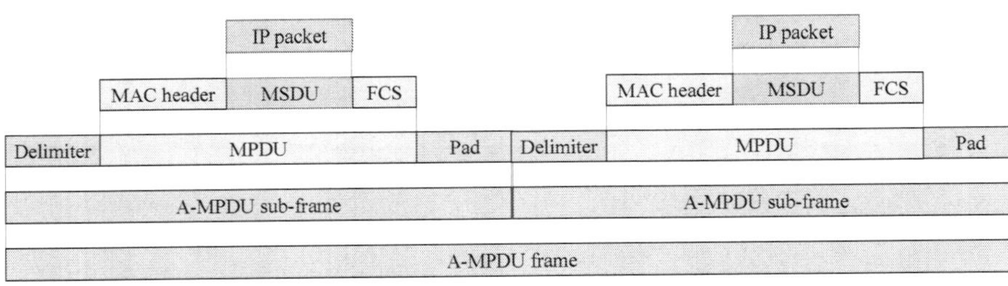

Figure 4.2. *Structure of A-MPDU frame*

With the exception of the last A-MPDU sub-frame, the padding bytes are added so that the size of each A-MPDU sub-frame is a multiple of four bytes.

The delimiter contains the size of the MPDU frame, an error check (CRC) on the frame size and a signature that can be used to detect a delimiter. The unique pattern is set to 4E in hexadecimal notation.

As each A-MPDU sub-frame gets its own MAC header, the encryption is applied to each sub-frame independently. Since each A-MPDU sub-frame has its own error detection sequence, an error will only affect the A-MPDU sub-frame, and the other A-MPDU sub-frames can be recovered.

All A-MPDU sub-frames must have the same destination on the radio link. On the other hand, the destination or the source address of the MPDU frame may be different.

4.1.3.2. *A-MSDU frame*

The A-MSDU (Aggregate MAC Service Data Unit) frame is a sequence of A-MSDU sub-frames (Figure 4.3). Each A-MSDU sub-frame contains an A-MSDU header, an MSDU data unit and pad bytes.

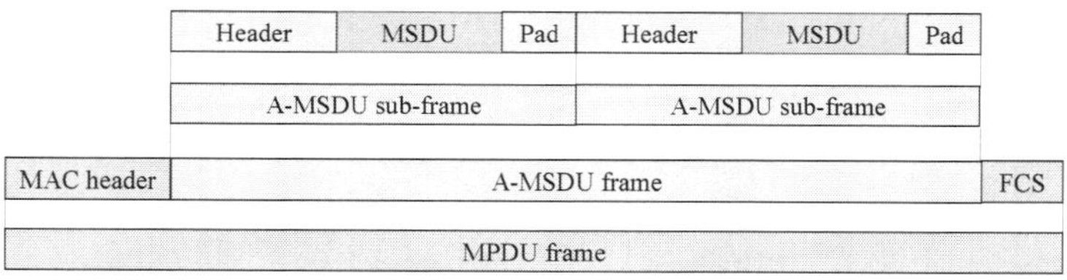

Figure 4.3. *Structure of A-MSDU frame*

The A-MSDU header consists of three fields: the MAC addresses of the destination and source of the MAC frame and the length of the MSDU data unit.

With the exception of the last A-MSDU sub-frame, padding bytes are added so that the size of each A-MSDU sub-frame is a multiple of four bytes.

Since the A-MSDU sub-frames of the same A-MSDU frame are contained in the same MPDU data unit, the same encryption applies to all the sub-frames.

Both forms of aggregation may be combined: an A-MPDU frame may contain an A-MSDU frame.

4.1.4. *Control frames*

4.1.4.1. *Block acknowledgment*

The block acknowledgment mechanism improves the efficiency of the channel by grouping multiple acknowledgments in a single control frame. There are two types of mechanisms: immediate acknowledgment and delayed acknowledgment.

The immediate acknowledgment mechanism is suitable for high-bandwidth and low-latency applications, whereas the delayed acknowledgment mechanism is suitable for applications that tolerate moderate latency.

The original design of the acknowledgment mechanism requires that each transmitted frame be acknowledged separately by the Ack control frame (Figure 4.4).

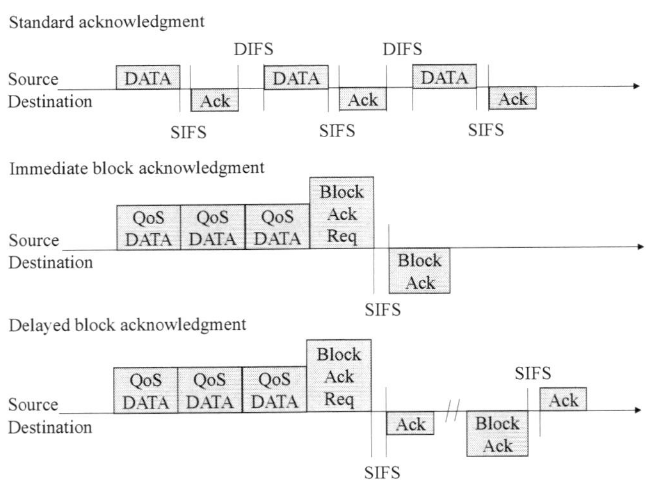

Figure 4.4. *Block acknowledgment*

If the immediate acknowledgment mechanism is used, then the recipient must respond to a BlockAckReq frame with a BlockAck frame (Figure 4.4). When the recipient sends the BlockAck frame, the initiator retransmits all frames that are not acknowledged in the BlockAck frame, either in another block or individually.

If the delayed acknowledgment mechanism is used, then the recipient must respond to a BlockAckReq control frame with an Ack control frame. The recipient must then send his response in a BlockAck control frame, which the initiator acknowledges by an Ack control frame (Figure 4.4).

The block acknowledgment mechanism has been introduced with the QoS mechanism. The block acknowledgment mechanism was initially optional, but the efficiency gains coupled with the aggregate frame transmission resulted in BlockAck control frame support being required for the 802.11n interface.

The initial definition of the block acknowledgment mechanism took into account the processing of frame-related sequence numbers and fragment numbers. For the 802.11n interface, the block acknowledgment mechanism can be compressed, thus only processing the sequence number.

4.1.4.2. *Control frame structure*

The BlockAckReq control frame is transmitted by the source of several MAC frames for block acknowledgment (Figure 4.5).

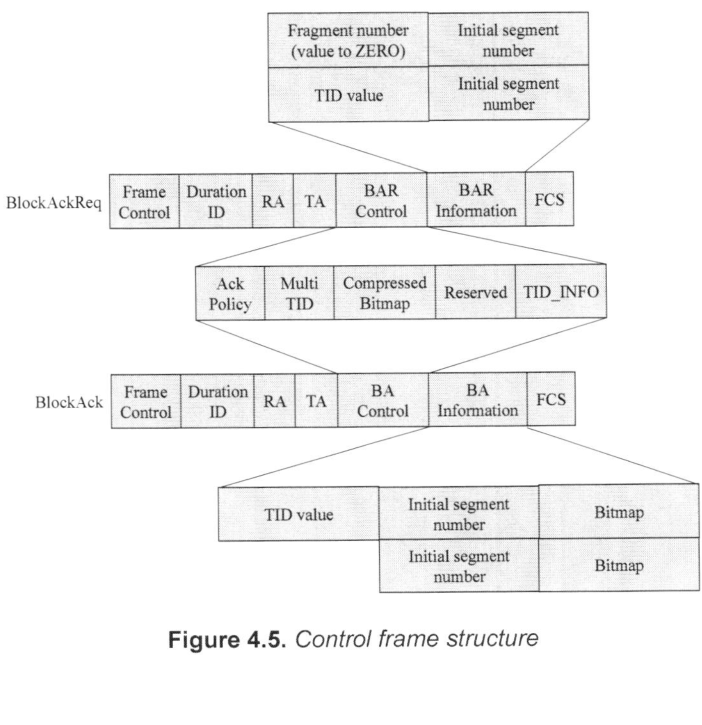

Figure 4.5. *Control frame structure*

The BAR or BA Control field contains the following information:

– Ack Policy: this bit indicates whether the block acknowledgment is immediate (bit to ZERO) or not (bit to ONE);

– Multi-TID: this bit indicates whether the block acknowledgment applies to several priority levels (bit to ONE) or not (bit to ZERO);

– Compressed Bitmap: this bit indicates whether the block acknowledgment is compressed (bit to ONE) or not (bit to ZERO);

– TID_INFO: this four-bit coded subfield provides the value of the TID (Traffic Identifier) field, for which a block acknowledgment is required.

The BAR or BA Information field contains the value of the sequence number of the first transmitted MAC frame. If the block acknowledgment applies to several priority levels, the sequence number is indicated for each priority level.

The BlockAck control frame is transmitted by the recipient of the MAC frames for block acknowledgment. Each bit in the bitmap of the Information field acknowledges (bit to ONE) or not (bit to ZERO) the frame that has this offset from the initial sequence number.

4.2. PLCP sub-layer

The PLCP (Physical Layer Convergence Procedure) sub-layer supports the following three modes:

– NON_HT mode: the PLCP header is identical to that defined for the 802.11a/g interfaces (Figure 4.6). NON_HT mode support is required;

– HT_MF (Mixed Format) mode: the PLCP header contains a preamble compatible with that defined for the 802.11a/g interfaces so that it can be processed by mobiles that do not handle the 802.11n interface (Figure 4.6). HT_MF mode support is required;

– HT_GF (Greenfield) mode: the PLCP header does not contain fields compatible with the 802.11a/g interfaces (Figure 4.6). Support for HT_GF format is optional. A mobile that does not support the HT_GF mode must be able to detect that a transmission of an HT_GF frame is in progress.

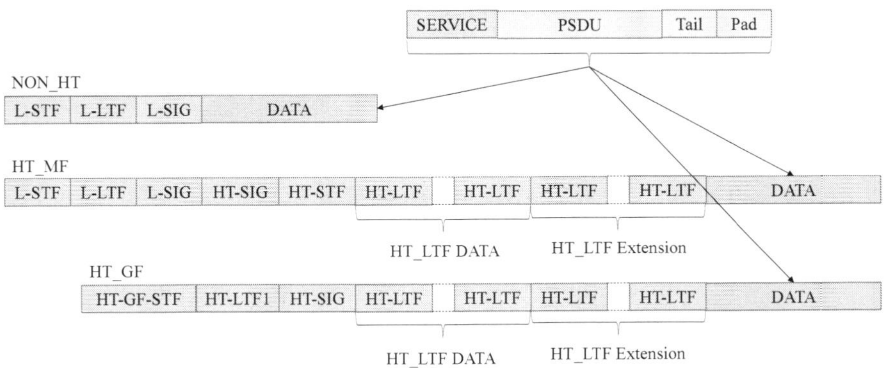

Figure 4.6. *PLCP frame structure*

L-STF (Non-HT Short Training Field): this field is identical to the short training sequence of the 802.11a/g interfaces.

L-LTF (Non-HT Long Training Field): this field is identical to the long training sequence of the 802.11a / g interfaces.

L-SIG (Non-HT Signal): this field is identical to the SIGNAL field of the 802.11a/g interfaces. This field allows mobiles that do not handle the 802.11n interface to determine the radio channel occupancy time, the bit rate being 6 Mbps.

HT-SIG (HT Signal): Table 4.4 describes the information in this field.

Information	Bit number	Designation
MCS	7	Index of the modulation and coding scheme (76 values)
Channel Bandwidth	1	Bit to ZERO for a bandwidth at 20 MHz Bit at ONE for a bandwidth at 40 MHz
HT Length	16	Size in bytes of the PSDU payload
Smoothing	1	Bit to ONE for smoothing the assessment of channel aggregation. Bit to ZERO for independent assessment of each channel
Not Sounding	1	Bit to ZERO if the PPDU data unit is a sounding Bit to ONE if not
Aggregation	1	Bit to ONE if the data unit contains A-MPDU sub-frames ZERO bit if not
STBC (Space-Time Block Coding)	2	2 bits to ZERO if transmission diversity is not used If not, value indicating the difference between the number of spatial/temporal diversity streams and the number of spatial flows
FEC (Forward Error Correction)	1	Bit to ZERO for LDPC (Low-Density Parity Check) coding Bit to UN for BCC (Binary Convolutional Code) coding
Short GI (Guard Interval)	1	Bit to ONE if a short guard interval is used ZERO bit if not
Number of extension spatial streams	2	Indicates the number of extension spatial streams. Set to 0 for no extension spatial stream. Set to 1 for 1 extension spatial stream. Set to 2 for 2 extension spatial streams. Set to 3 for 3 extension spatial streams.
CRC	8	Cyclic redundancy code
Tail	6	Tail of the convolutional encoder

Table 4.4. *HT-SIG field structure*

HT-STF (HT Short Training Field): this field has the same purpose as the L-STF field.

There are two types of HT-LTF (HT Long Training Field):

– DATA HT-LTF field helps to set the MIMO (Multiple Input Multiple Output) mechanism;

– HT-LTF Extension field is used for beamforming. The number of HTF LTF fields depends on the number of spatial flows. This field is optional.

4.3. PMD sub-layer

Table 4.5 summarizes the features provided by the PMD (Physical Medium Dependent) sub-layer.

Features	Mandatory/optional
BPSK, QPSK, 16QAM, 64QAM Modulation	Mandatory
BCC error correction code	Mandatory
LDPC error correction code	Optional
Short guard interval (400 ns)	Optional
MIMO (up to four streams)	Optional
Beamforming	Optional
STBC	Optional

Table 4.5. *Characteristics of PMD sub-layer*

4.3.1. *Transmission chain*

Transmission in the HT_MF and HT_GF modes is generated from the following function blocks:

a) Scrambler scrambles the data to reduce the probability of long sequences of bits to ZERO or to ONE.

b) Encoder parser, if BCC encoding is to be used, demultiplexes the scrambled bits among N_{ES} (number of BCC encoders for the Data field) BCC encoders, in a round robin manner.

c) FEC encoders encode the data to enable error correction. An FEC encoder may include a binary convolutional encoder followed by a puncturing device, or it may include an LDPC encoder.

d) Stream parser divides the outputs of the encoders into blocks that are sent to different interleaver and mapping devices. The sequence of the bits sent to an interleaver is called a spatial stream.

e) Interleaver interleaves the bits of each spatial stream (changes order of bits) to prevent long sequences of adjacent noisy bits from entering the BCC decoder. Interleaving is applied only when BCC encoding is used.

f) Constellation mapper maps the sequence of bits in each spatial stream to constellation points (complex numbers).

g) STBC encoder spreads constellation points from N_{SS} spatial streams into N_{STS} space-time streams using a space-time block code. STBC is used only when $N_{SS}<N_{STS}$.

h) Spatial mapper maps space-time streams to transmit chains. This may include one of the following:

– direct mapping: constellation points from each space-time stream are mapped directly onto the transmit chains (one-to-one mapping);

– spatial expansion: vectors of constellation points from all the space-time streams are expanded via matrix multiplication to produce the input to all the transmit chains;

– beamforming: similar to spatial expansion, each vector of constellation points from all the space-time streams is multiplied by a matrix of steering vectors to produce the input to the transmit chains.

i) Inverse discrete Fourier transform (IDFT) converts a block of constellation points to a time domain block.

j) CSD (Cyclic Shift Diversity) insertion is where the insertion of the cyclic shifts prevents unintentional beamforming. CSD insertion may occur before or after the IDFT.

k) GI insertion prepends to the symbol a circular extension of itself.

l) Windowing optionally smooths the edges of each symbol to increase spectral decay.

Figure 4.7 shows the blocks used to generate the HT-SIG field of the PPDU data unit in HT_MF mode.

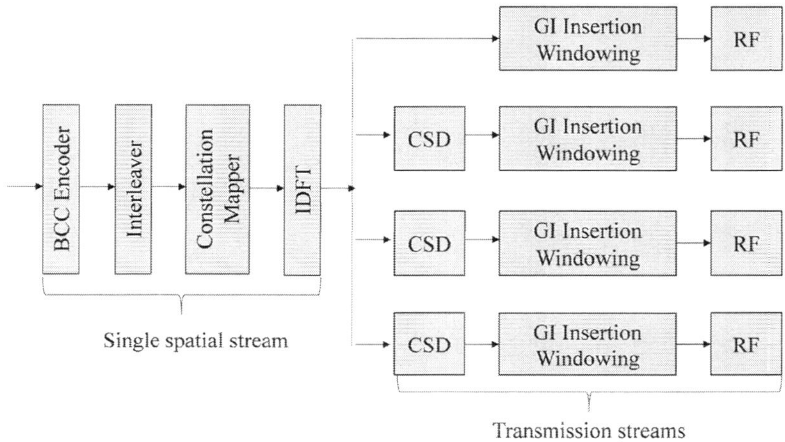

Figure 4.7. *Transmission chain – Diagram 1*

These blocks are also used to generate the NON_HT part of the PPDU data unit in HT_MF mode.

The BCC encoder and the interleaver are not used when generating the L-STF and L-LTF fields.

Figure 4.8 shows the blocks used to generate the DATA field for HT_MF and HT_GF modes.

A subset of these blocks consisting of the constellation mapper and the CSD blocks, as well as the blocks on the right, including the spatial mapping block, is also used to generate the HT-STF, HT-GF-STF and HT-LTF fields.

The HT-GF-SIG field is generated using the blocks shown in Figure 4.7, augmented by additional CSD blocks and spatial mapping.

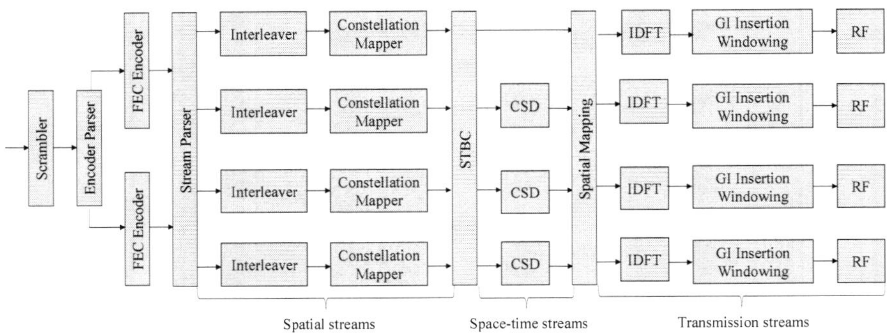

Figure 4.8. *Transmission chain – Diagram 2*

4.3.2. *Frequency plan*

The 802.11n interface operates in the N-NII (Unlicensed-National Information Infrastructure) band, at 5 GHz, as the 802.11a interface, and in the ISM (Industrial, Scientific and Medical) band, at 2.4 GHz, as the 802.11g interface.

The 802.11n interface uses the 20 MHz radio channel, as for the 802.11a/g interfaces, and offers the possibility of aggregating two adjacent radio channels in the U-NII band at 5 GHz (Figure 4.9).

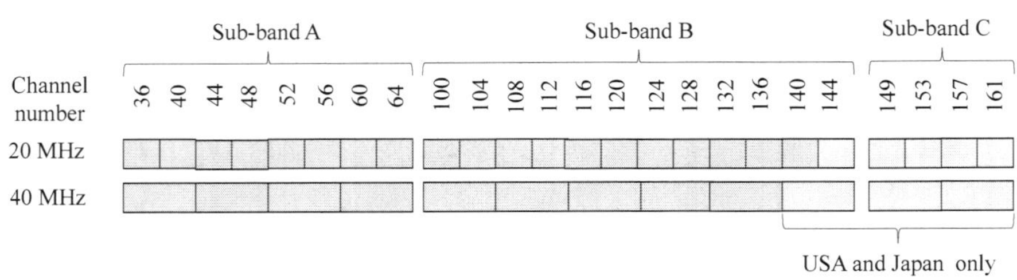

Figure 4.9. *Frequency plan*

4.3.3. *Frequency multiplexing*

For the 20 MHz radio channel band, for HT modes, the complex numbers are numbered 0 to 51 and are subsequently mapped onto OFDM (Orthogonal Frequency-Division Multiplexing) sub-carriers, numbered −28 to −22, −20 to −8, −6 to −1, 1 to 6, 8 to 20 and 22 to 28 (Table 4.6).

Sub-carriers −21, −7, 7 and 21 are ignored and subsequently used for the insertion of pilot sub-carriers.

Parameters	NON_HT	HT 20 MHz	HT 40 MHz
N_{SD} Number of sub-carriers assigned to the DATA field	48	52	108
N_{SP} Number of sub-carriers assigned to pilots	4	4	6
N_{ST} Total number of sub-carriers	52	56	114
Δ_F Spacing between sub-carriers	0.3125 MHz (20 MHz/64)		

Table 4.6. *OFDM multiplexing parameters*

For the 40 MHz radio channel band, the complex numbers are numbered 0 to 107 and are subsequently mapped onto OFDM sub-carriers numbered −57 to −54, −52 to −26, −24 to −12, −10 to −1, 1 to 10, 12 to 24, 26 to 52 and 54 to 57 (Table 4.6).

Sub-carriers −53, −25, −11, 11, 25 and 53 are ignored and subsequently used for the insertion of pilot sub-carriers.

Sub-carrier 0, associated with the central frequency, is omitted and filled with the value of ZERO.

4.3.4. *Space multiplexing*

4.3.4.1. *MIMO mechanism*

The MIMO mechanism consists of simultaneously transmitting m signals and receiving them on n antennas, with m <n, using the same radio channel. Each receiving antenna receives the m transmitted signals, each signal being modified by the transfer function between the transmitting and receiving antennas. There is thus a transmission matrix H of size m × n (Figure. 4.10).

The MIMO mechanism, by spatially multiplexing m signals, makes it possible to increase the rate of the radio channel with the same factor.

The MIMO mechanism uses the transmission matrix H to perform spatial demultiplexing.

For the SU (Single User) MIMO mechanism, the m transmitted signals are destined for the same user.

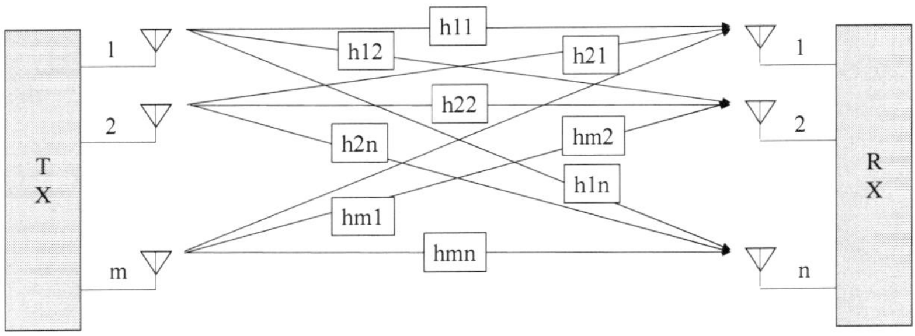

Figure 4.10. *MIMO mechanism*

4.3.4.2. *STBC mechanism*

When the number of transmitters (m) is greater than the number of receivers (n), the additional transmitters are used to effect transmit diversity, thereby improving the quality of the received signal by protecting the transmission from fading.

The STBC mechanism performs space and time diversity. The signal S corresponding to a spatial flux is divided into two parts, S1 and S2. The complex numbers of the S (= S1 + S2) constellation of N_{SS} spatial streams (N_{SS} = 1) are distributed over N_{STS} space-time flows (N_{STS} = 2) (Figure 4.11).

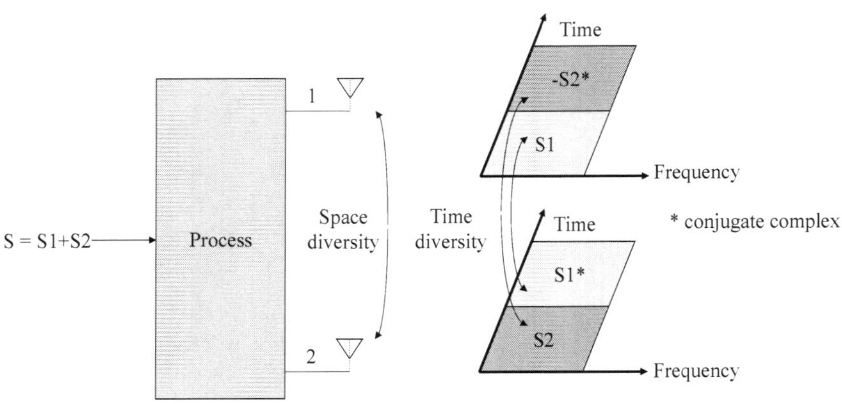

Figure 4.11. *STBC mechanism*

4.3.4.3. *Beamforming*

Beamforming allows a transmitter, called a beamformer, to focus the energy of several sources of the same signal in the direction of the receiver, called a beamformee. Phase reception increases the signal-to-noise ratio of the received signal (Figure 4.12).

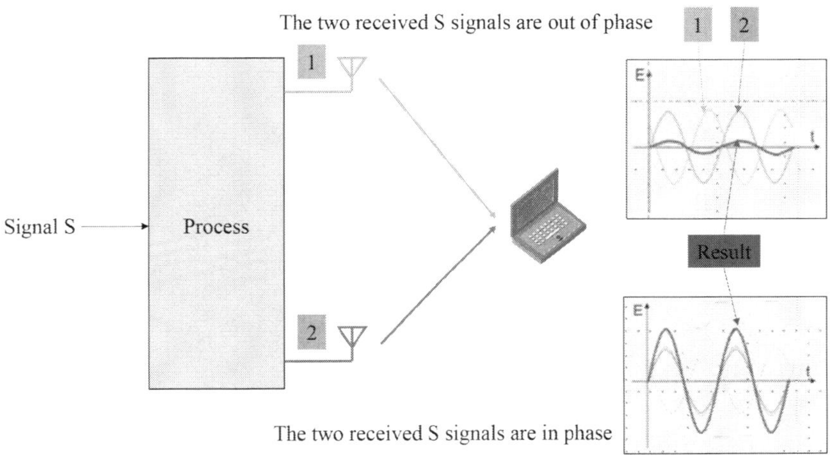

Figure 4.12. *Beamforming mechanism*

For the explicit beamforming, a device measures the radio channel and uses this measurement to directly calculate the direction matrix. The active channel measurement is performed by transmitting a sounding to the beamformee, which responds with a frame that indicates how the sounding was received.

For implicit beamforming, frames such as ACK control frames or data transmitted on pilot channels can be used to estimate the direction matrix.

4.3.5. *Modulation and coding scheme*

The value of the modulation and coding scheme (MCS) determines the rate value from the following parameters:

– the modulation of the sub-carriers, phase modulation (BPSK or QPSK) or mixed phase and amplitude modulation (16-QAM or 64-QAM);

– the coding rate of the error correction code, which can take values of 1/2, 2/3, 3/4 or 5/6;

– the number of spatial flows of the MIMO mechanism, which can take values of 1, 2, 3 or 4;

– the bandwidth of the radio channel, which may be 20 or 40 MHz;

– the duration of the guard interval, short duration of 400 ns or long duration of 800 ns.

Table 4.7 (respectively Table 4.8) provides the rate values, for a single spatial stream, for the first eight MCS values (between 0 and 7), for a bandwidth of 20 MHz (respectively 40 MHz).

Modulation	Coding rate	Number of bits per sub-carrier	Number of DATA bits per OFDM symbol	Number of encoded DATA bits	Rate (Mbps)	
					GI 800 ns	GI 400 ns
BPSK	1/2	1	26	52	6.5	7.2
QPSK	1/2	2	52	104	13.0	14.4
QPSK	3/4	2	78	104	19.5	21.7
16-QAM	1/2	4	104	208	26.0	28.9
16-QAM	3/4	4	156	208	39.0	43.3
64-QAM	2/3	6	208	312	52.0	57.8
64 QAM	3/4	6	234	312	58.5	65.0
64-QAM	5/6	6	260	312	65.0	72.2

Table 4.7. *Parameters of the modulation and coding scheme 20 MHz bandwidth*

Modulation	Coding rate	Number of bits per sub-carrier	Number of DATA bits per OFDM symbol	Number of encoded DATA bits	Rate (Mbps)	
					GI 800 ns	GI 400 ns
BPSK	1/2	1	54	108	13.5	15.0
QPSK	1/2	2	108	216	27.0	30.0
QPSK	3/4	2	162	216	40.5	45.0
16-QAM	1/2	4	216	432	54.0	60.0
16-QAM	3/4	4	324	432	81.0	90.0
64-QAM	2/3	6	432	648	108.0	120.0
64 QAM	3/4	6	486	648	121.5	135.0
64-QAM	5/6	6	540	648	135.0	150.0

Table 4.8. *Parameters of the modulation and coding scheme 40 MHz bandwidth*

The MCS values between 8 and 15 are relative to two spatial streams, for a bandwidth of 20 or 40 MHz.

The MCS values between 16 and 23 are relative to three spatial streams, for a bandwidth of 20 or 40 MHz.

The MCS values between 24 and 31 are relative to four spatial streams, for a bandwidth of 20 or 40 MHz.

The MCS values between 33 and 76 correspond to the STBC mechanism for which:

– two spatial flows (N_{SS}=2) are coded in three space-time flows (N_{STS} = 3);

– three spatial flows (N_{SS}=3) are coded in four space-time flows (N_{STS} = 4).

MCS values between 0 and 15, with a guard interval of 800 ns and a bandwidth of 20 MHz are required. Other MCS values, 400 ns guard interval and 40 MHz bandwidth are optional.

The MCS value of 32 has the characteristics described in Table 4.9.

Modulation	Coding rate	Number of bits per sub-carrier	Number of DATA bits per OFDM symbol	Number of encoded DATA bits	Rate (Mbps)	
					GI 800 ns	GI 400 ns
BPSK	1/2	1	24	48	6.0	6.7

Table 4.9. *MCS 32 parameters*

5

802.11ac Interface

5.1. MAC layer

5.1.1. *Management frame evolution*

5.1.1.1. *VHT Capabilities information element*

The management frames indicate that the access point has an 802.11ac interface by including the VHT (Very High Throughput) Capabilities information element, containing the following fields.

The information provided by the VHT Capabilities Info field is described in Table 5.1.

Information	Designation
Maximum MPDU Length	Indication of the maximum size of the MPDU frame (3,895, 7,991 or 11,454 bytes)
Supported Channel Width Set	Indication of the bandwidth of the radio channel (80 + 80 and/or 160 MHz)
Rx LDPC	Using the LDCP error correction code on reception
Short GI for 80 MHz	Short guard interval for the 80 MHz radio channel
Short GI for 160 and 80+80 MHz	Short guard interval for 80 + 80 or 160 MHz radio channel
Tx STBC	Transmission for the space-time diversity STBC

Rx STBC	Indication of the number of spatial streams supported for STBC diversity
SU Beamformer Capable	Indication relating to the beamforming for a single user, at the beamformer level
SU Beamformee Capable	Indication relating to the beamforming for a single user, at the beamformee level
Beamformee STS Capability	Indication of the maximum number of space-time stream supported for beamforming by the beamformee
Number of Sounding Dimensions	Number of antennas involved in the assessment of the radio channel
MU Beamformer Capable	Indication relating to the beamforming for several mobiles, at the beamformer level
MU Beamformee Capable	Indication relating to the beamforming for several mobiles, at the beamformee level
VHT TXOP PS	Use of power-saving
+HTC-VHT Capable	Ability to receive frames containing the variant HT Control field
Maximum A-MPDU Length Exponent	Indication about the maximum size of the A-MPDU frame that the access point can handle on reception
VHT Link Adaptation Capable	Use of information Link Adaptation of the variant HT Control field
Receive Antenna Pattern Consistency	Indication of a possible change in antenna pattern for reception during the association period
Transmit Antenna Pattern Consistency	Indication of a possible change in antenna pattern for transmission during the association period

Table 5.1. *Subfields of the VHT Capabilities Info field*

The field-supported VHT-MCS and NSS Set is used to transmit the combinations of modulation and coding schemes and spatial streams that an access point supports for transmission and reception.

5.1.1.2. *VHT Operation information element*

The mobile obtains the information on the primary radio channel from the HT Operation information element.

The VHT Operation information element provides mobiles with the additional features of the 802.11ac interface and contains the following fields:

Channel Width: this field indicates the bandwidth of the radio channel (20, 40, 80, 160 or 80 + 80 MHz).

Channel Center Frequency Segment 0: this field indicates the radio channel center for 80 or 160 MHz bandwidths. For the bandwidth of 80 + 80 MHz, this field indicates the center of the 80 MHz radio channel of segment 0.

Channel Center Frequency Segment 1: this field indicates the center of the 80 MHz radio channel of segment 1, for the 80 + 80 MHz frequency band.

5.1.1.3. *Extended BSS Load information element*

The Extended BSS Load information element contains information about the underutilization of the MIMO spatial stream and the use of bandwidth. A mobile receiving this piece of information can use it in a specific access point selection algorithm.

The Extended BSS Load information element contains the following fields:

MU-MIMO Capable STA Count: this field indicates the total number of mobiles associated with the access point whose MU Beamformee Capable field of the VHT Capabilities information element is set to ONE.

Spatial Stream Underutilization: this field is defined as the percentage of time underutilized by the access point in the spatial domain, for a busy time of the radio channel. Underutilization is calculated only for the primary channel.

Observable Secondary 20 MHz Utilization, Observable Secondary 40 MHz Utilization and Observable Secondary 80 MHz Utilization: these fields indicate the load of the secondary radio channels, in conjunction with the measurement of the primary channel.

5.1.1.4. *Wide Bandwidth Channel Switch information element*

The Wide Bandwidth Channel Switch information element is used to switch to a new radio channel. It contains the New Channel Width, New Channel Center Frequency Segment 0 and New Channel Center Frequency Segment 1 fields with the same definition as the VHT Operation information element.

5.1.1.5. *Channel Switch Wrapper information element*

The Channel Switch Wrapper information element indicates the characteristics of the cell after switching the radio channel.

The Channel Switch Wrapper information element contains the following fields:

The New Country field contains the necessary information to enable a mobile to identify the new domain in which it is located.

The Wide Bandwidth Channel Switch field is similar to the eponymous information element.

5.1.1.6. *VHT Transmit Power Envelope information element*

The VHT Transmit Power Envelope information element transmits the maximum transmit power for the various bandwidths of the radio channel.

5.1.1.7. *Quiet Channel information element*

The Quiet Channel information element indicates the time interval during which the secondary 80 MHz radio channel must be silent. It also indicates the conditions of use of the 80 MHz primary channel.

5.1.1.8. *Operating Mode Notification information element*

The Operating Mode Notification information element indicates that the access point is modifying the radio channel width and the maximum number of spatial streams it can receive.

5.1.2. *Control frames*

RTS (Request To Send) and CTS (Clear To Send) control frames are used to negotiate the value of the available bandwidth.

The TA (Transmitter Address) field of the RTS control frame may contain information relating to the bandwidth of the radio channel.

The RA field of the CTS control frame copies the value of the TA field of the RTS control frame if the required bandwidth is available. In the opposite case, the value of the RA field indicates the value of the bandwidth actually available.

When an 802.11ac device sends an RTS control frame, it must verify that the bandwidth (80 MHz) of the radio channel is available. The control frame is transmitted in a 20 MHz radio channel and replicated to fill the 80 MHz radio channel (Figure 5.1).

Each device, whether it is 802.11a/n/ac equipment, receives the RTS control frame and interprets the Duration field to determine the occupancy time of the 20 MHz radio channel.

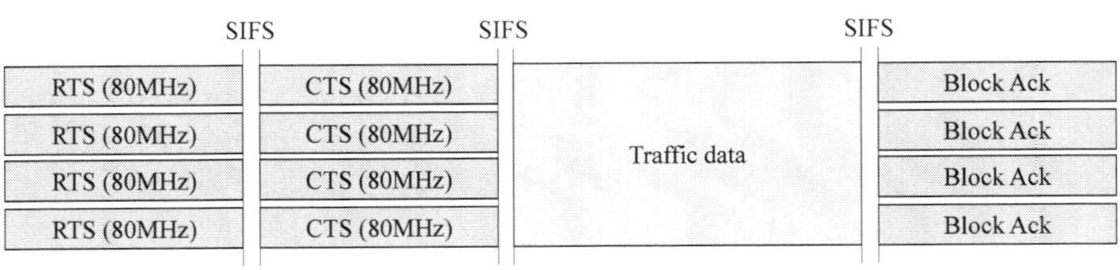

a) The bandwidth of 80 MHz is available

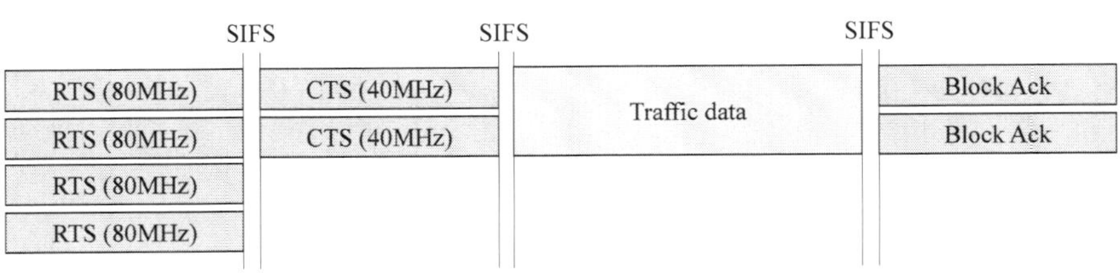

b) The bandwidth of 40 MHz is available

Figure 5.1. *Bandwidth negotiation*

Before the equipment addressed by the RTS control frame (the receiver) responds with the CTS control frame, it checks whether a transmission is taking place on its main radio channel or any other radio channel at 20 MHz in the band of 80 MHz (Figure 5.1).

If the bandwidth of 80 MHz is free, then the recipient reports it with four duplicate CTS control frames.

If a portion of the bandwidth is used, the recipient responds with a CTS control frame only on the available 20 MHz radio channels indicating the overall bandwidth (e.g. 40 MHz).

5.1.3. *MAC header structure*

The 802.11ac interface maintains the structure of the MAC header defined for the 802.11n interface. The HT Control field has two forms: the HT variant and the VHT variant. These two forms differ according to the HT Control Middle field format. For the HT variant, the first bit is ZERO, whereas for the VHT variant, the HT Control field starts with a bit at ONE.

The MAC header, described in Figure 5.2, has the following fields.

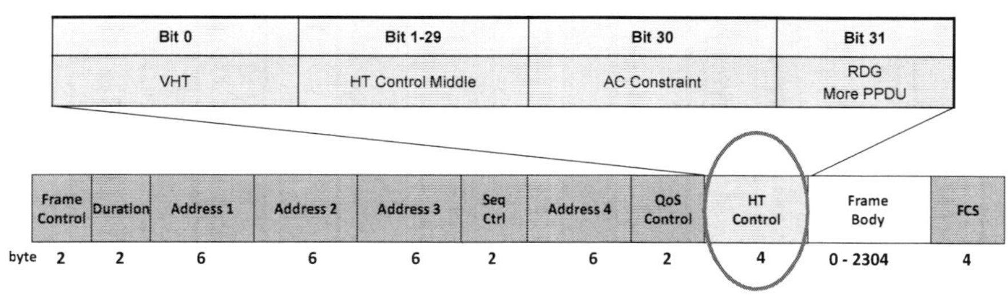

Figure 5.2. *MAC header structure*

The information provided by the HT Control Middle field is described in Table 5.2.

Information	Designation
MRQ (VHT-MCS feedback Request)	Request regarding the recommended value of the modulation and coding scheme
MSI (MRQ Sequence Identifier) STBC indication	MSI: MRQ request identifier STBC: indicates whether the estimate in the MFB subfield is calculated based on a PPDU data unit using STBC
MFSI (MCS Feedback Sequence Identifier) GID-L (see note)	MFSI: identifier of the sequence relating to a request for a recommendation on the value of the modulation and coding scheme (MCS) GID-L: part of the group identifier (GID) the PPDU data unit used to determine the value of the modulation and coding scheme indicated in the MFB field
MFB	Recommended modulation and coding scheme (MCS)
GID-H	Complementary part of the group identifier (GID) of the PPDU data unit
Coding Type	Type of error correction code BCC (Binary Convolutional Code) LDPC (Low-Density Parity Check)
FB Tx Type	Transmission type of the measured PPDU data unit
Unsolicited MFB	Indication of unsolicited information relating to a recommendation on the value of the modulation and coding scheme

Note: the group identifier (GID) is assigned to the mobile in the case of a MU-MIMO (Multi-User – Multiple Input Multiple Output) transmission

Table 5.2. *Subfields of Control Middle field*

5.2. PLCP sub-layer

The structure of the PLCP (Physical Layer Convergence Procedure) sub-layer is described in Figure 5.3.

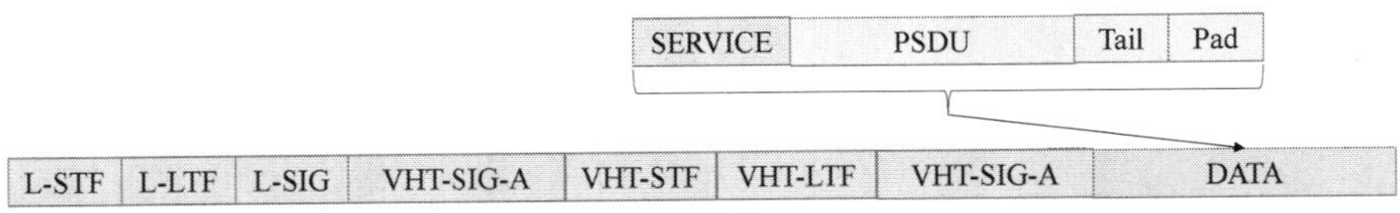

Figure 5.3. *PLCP frame structure*

L-STF (Non-HT Short Training Field): this field is identical to the short training sequence of the 802.11a/g/n interfaces.

L-LTF (Non-HT Long Training Field): this field is identical to the long training sequence of the 802.11a/g/n interfaces.

The VHT-SIG-A, VHT-STF, VHT-LTF and VHT-SIG-B fields only exist for the VHT format.

VHT-SIG-A: this field is used to interpret the PPDU data unit. Table 5.3 describes the information in this field.

The main purpose of the VHT-STF field is to improve the estimation of automatic gain control for MIMO transmission.

The VHT-LTF field allows the receiver to estimate the transfer function of the radio channel for MIMO transmission.

The VHT-SIG-B field provides information on the length of the payload of the PPDU data unit (for SU-MIMO or MU-MIMO transmission) and on the modulation and coding scheme (only for MU-MIMO transmission).

Information	Bit number	Designation
BW	2	Bandwidth of the radio channel (20, 40, 80, 160 or 80+80 MHz)
STBC (Space-Time Block Coding)	1	Use of transmission diversity
Group ID	6	Identifier assigned to a mobile set in the case of a MU-MIMO transmission
NSTS Partial AID	12	NSTS: number of space-time channels Partial AID: part of the mobile identifier
TXOP_PS_ NOT_ALLOWED	1	Entering the power-saving mode
Short GI (Guard Interval)	1	Bit to ONE if a short guard interval is used Bit to ZERO if not
Short GI / NSYM Disambiguation	1	Bit to ONE if a short guard interval is used and if the number of symbols mod10 = 9 Bit to ZERO if not
SU/MU[0] Coding	1	Bit to ONE for LDPC (Low-Density Parity Check) encoder Bit to ZERO for BCC (Binary Convolutional Code) encoder
LDPC Extra OFDM Symbol	1	In the case of LDPC encoder, indication of an additional symbol used
SU VHT MCS/ MU[1-3] Coding	4	Modulation and coding scheme (MCS)
Beamformed	1	Applying a direction matrix for the beamforming
CRC	8	Cyclic Redundancy Check
Tail	6	Completion of the lattice of the convolution code

Table 5.3. *Structure of VHT-SIG-A field*

5.3. PMD sub-layer

The PMD (Physical Medium Dependent) sub-layer improves the rate of the 802.11ac radio interface relative to the 802.11n interface from the following parameters:

– aggregation of radio channels makes it possible to constitute a bandwidth of 80, 160 or 80 + 80 MHz. For the 802.11n interface, aggregation limits the bandwidth to 40 MHz.

– 256-QAM modulation allows an OFDM symbol to carry eight bits. For the 802.11n interface, 64-QAM modulation limits the number of bits to 6.

– spatial multiplexing MIMO can be performed on eight space-time flows. For the 802.11n interface, spatial multiplexing is limited to four space-time flows.

5.3.1. *Transmission chain*

The transmission chain includes the function blocks described for the 802.11n interface in section 4.3.1.

Figure 5.4 shows the transmission process for the L-SIG and VHT-SIG-A fields of a PPDU data unit. These transmit blocks are also used to generate the non-VHT fields of the VHT data unit.

Figures 5.5 and 5.6 show the transmission process for generating the VHT-SIG-B field of a PPDU data unit for a single user (SU) and multi-user (MU), for radio channel bandwidths of 20, 40 and 80 MHz.

Figures 5.7 and 5.8 show the transmission process for generating the VHT-SIG-B field of a PPDU data unit for a single user (SU), for radio channel bandwidths of 160 and 80 + 80 MHz.

Figures 5.9 and 5.10 show the transmission process for generating the DATA field of PPDU data unit for a single user, for radio channel bandwidths of 20, 40 and 80 MHz, with BCC and LDCP encoders. A subset of these transmit blocks is also used to generate the VHT-LTF and VHT-STF fields.

Figure 5.11 shows the transmission process for generating the DATA field of a PPDU data unit for multi-user (MU), for radio channel bandwidths of 20, 40 and 80 MHz, with BCC and LDPC encoders.

Figures 5.12 and 5.13 show the transmission process for generating the DATA field of a PPDU data unit for a single user (SU), for a radio channel bandwidth of 160 MHz, with BCC and LDPC encoders.

Figures 5.14 and 5.15 show the transmission process for generating the DATA field of a PPDU data unit for a single user (SU), for a radio channel bandwidth of 80 + 80 MHz, with BCC and LDPC encoders.

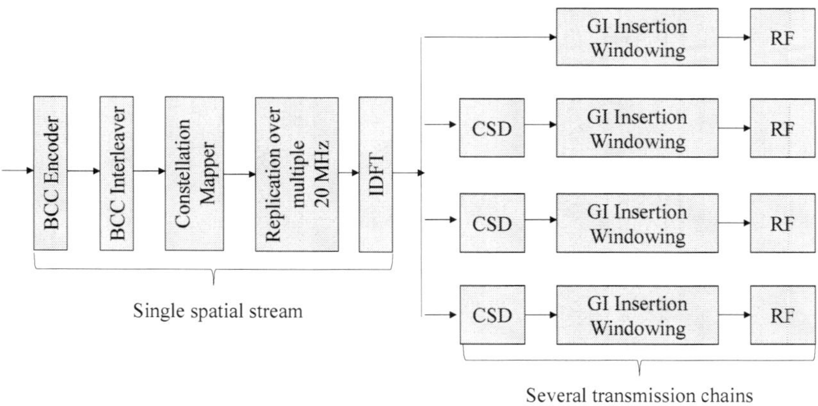

Figure 5.4. *Generation of L-SIG and VHT-SIG-A fields*

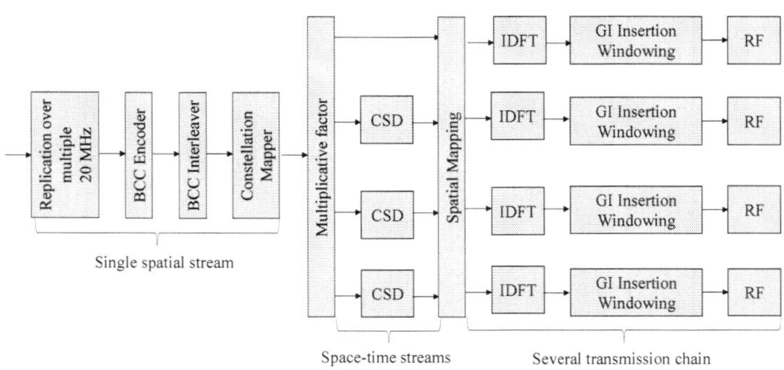

Figure 5.5. *Generation of VHT-SIG-B field – Data unit for a single user Radio channel bandwidths of 20, 40, and 80 MHz*

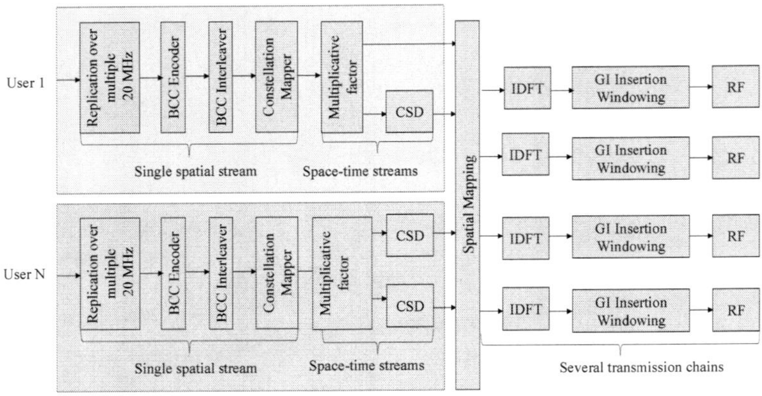

Figure 5.6. *Generation of VHT-SIG-B field – Data unit for multi-user Radio channel bandwidths of 20, 40, and 80 MHz*

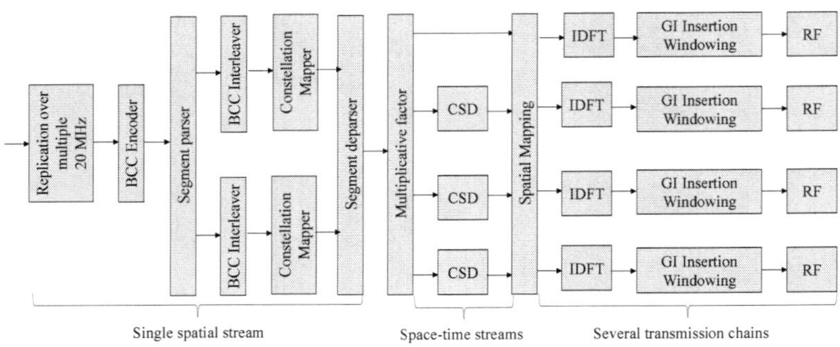

Figure 5.7. *Generation of VHT-SIG-B field – Data unit for a single user Radio channel bandwidth of 160 MHz*

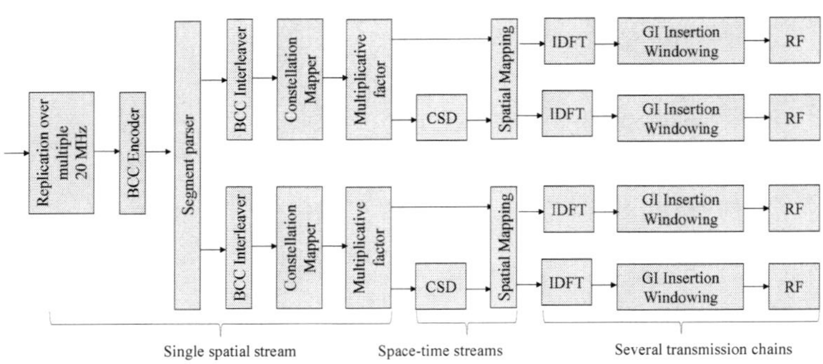

Figure 5.8. *Generation of VHT-SIG-B field – Data unit for a single user Radio channel bandwidth of 80+80 MHz*

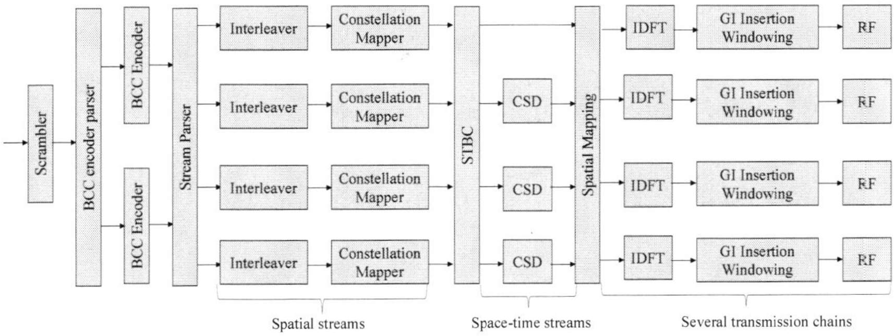

Figure 5.9. *Generation of DATA field – Data unit for a single user BCC encoder – Radio channel bandwidths of 20, 40 and 80 MHz radio*

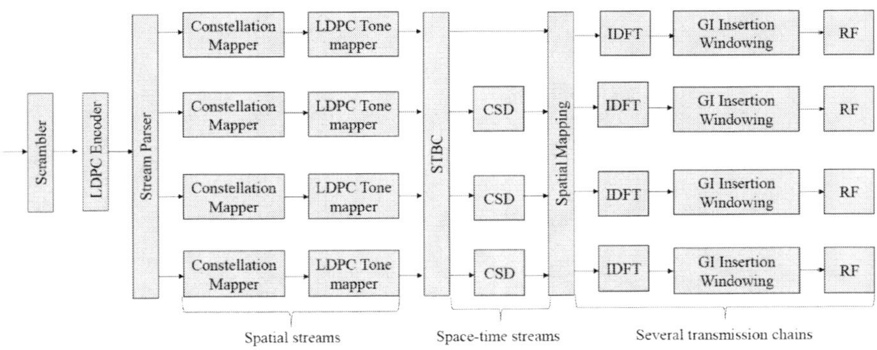

Figure 5.10. *Generation of DATA field – Data unit for a single user LDPC encoder – Radio channel bandwidths of 20, 40 and 80 MHz radio*

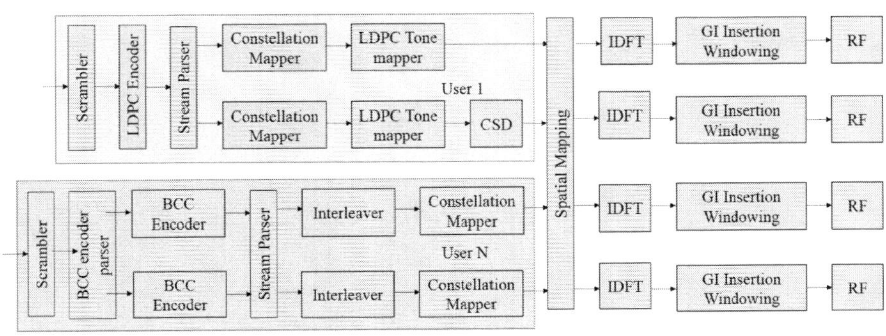

Figure 5.11. *Generation of DATA field – Data unit for multi-user Radio channel bandwidths of the of 20, 40 and 80 MHz*

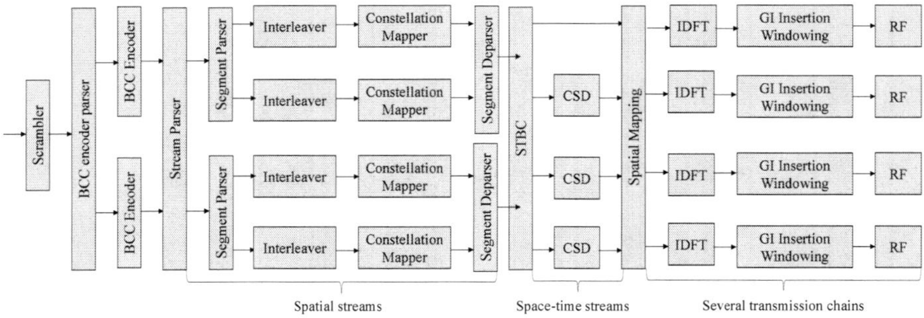

Figure 5.12. *Generation of DATA field – Data unit for a single user BCC encoder – Radio channel bandwidth of the 160 MHz*

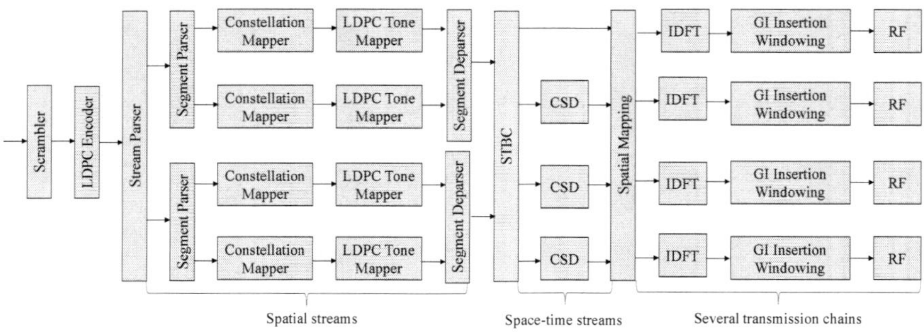

Figure 5.13. *Generation of DATA field – Data for a single user LDPC encoder – Radio channel bandwidth of the 160 MHz*

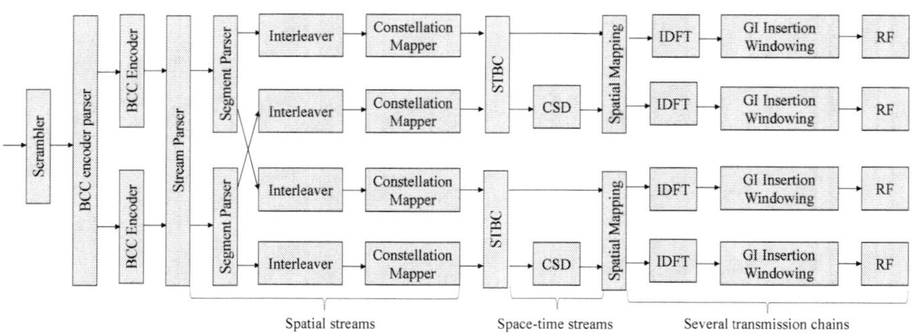

Figure 5.14. *Generation of DATA field – Data unit for a single user BCC encoder – Radio channel bandwidth of 80 + 80 MHz*

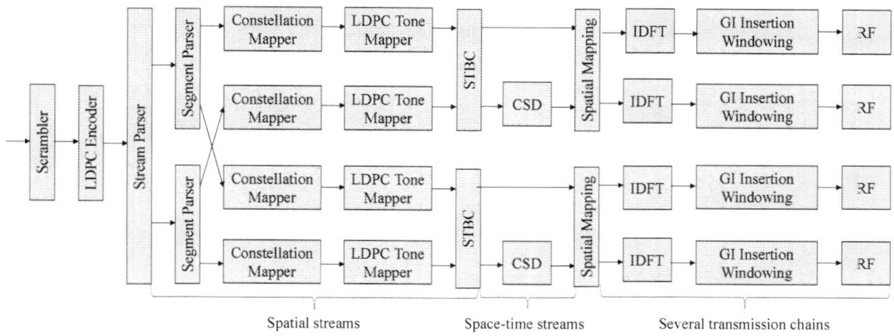

Figure 5.15. *Generation of DATA field – Data unit for a single user LDPC encoder – Radio channel bandwidth of 80 + 80 MHz*

5.3.2. *Frequency plan*

The 802.11ac interface operates in the U-NII (Unlicensed-National Information Infrastructure) band at 5 GHz.

Adjacent radio channels of 20 MHz are grouped in pairs to form a 40 MHz channel. Adjacent radio channels of 40 MHz are grouped in pairs to form an 80 MHz channel. Adjacent radio channels of 80 MHz are grouped in pairs to form a 160 MHz channel (Figure 5.16).

It is allowed for two 802.11ac access points to select the same 80 MHz bandwidth, but one access point must have its primary channel of 20 MHz in the lower half-band, and the other access point puts its primary channel of 20 MHz in the upper half-band.

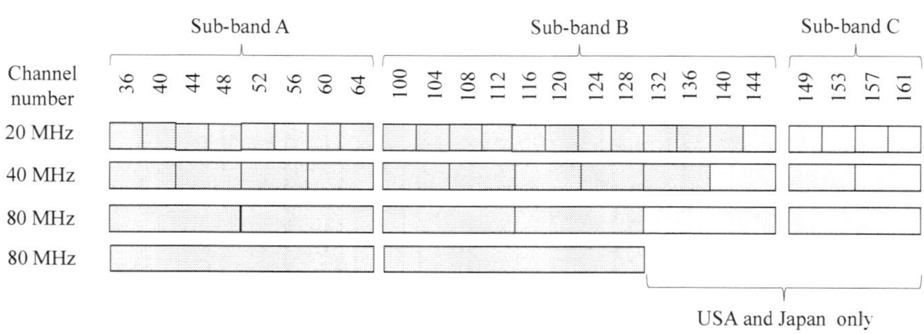

Figure 5.16. *Frequency plan Channel bandwidths of 20, 40, 80 and 160 MHz*

The 802.11n mobiles associated with the first access point can transmit in the lower half-band of 20 or 40 MHz, at the same time that the 802.11n mobiles associated with the second access point transmit in the upper half-band of 20 or 40 MHz. An 802.11ac mobile can transmit in the 80 MHz band if it is available.

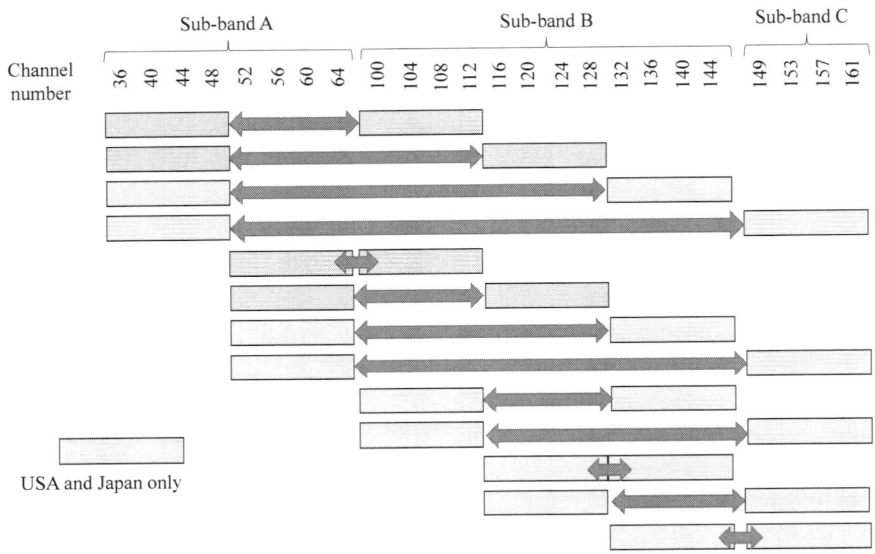

Figure 5.17. *Frequency plan Channel bandwidths of 80+80 MHz*

The frequency plan allows only two channels at 160 MHz. Each channel can be constrained by regulatory requirements. The aggregation of two non-contiguous channels of 80 MHz defines 13 possibilities, thus allowing greater flexibility for interference protection (Figure 5.17).

5.3.3. *Frequency multiplexing*

For a radio channel of 20 MHz, the bandwidth is divided into 64 sub-carriers. The signal is transmitted on the sub-carriers −28 to −1 and 1 to 28, with sub-carrier 0 being the central sub-carrier.

For a radio channel of 40 MHz, the bandwidth is divided into 128 sub-carriers. The signal is transmitted on the sub-carriers −58 to −2 and 2 to 58.

For a radio channel of 80 MHz, the bandwidth is divided into 256 sub-carriers. The signal is transmitted on the sub-carriers −122 to −2 and 2 to 122.

For a radio channel of 160 MHz, the bandwidth is divided into 512 sub-carriers. The signal is transmitted on the sub-carriers −250 to −130, −126 to −6, 6 to 126 and 130 to 250.

For a non-contiguous radio channel of 80 + 80 MHz, each frequency segment of 80 MHz is divided into 256 sub-carriers. In each frequency segment, the signal is transmitted on sub-carriers −122 to −2 and 2 to 122.

5.3.4. *Spatial multiplexing*

On the 802.11n interface, an access point can simultaneously transmit several streams spatially multiplexed to a single station using the SU-MIMO (Single User – Multiple Input Multiple Output) mechanism (Figure 5.18).

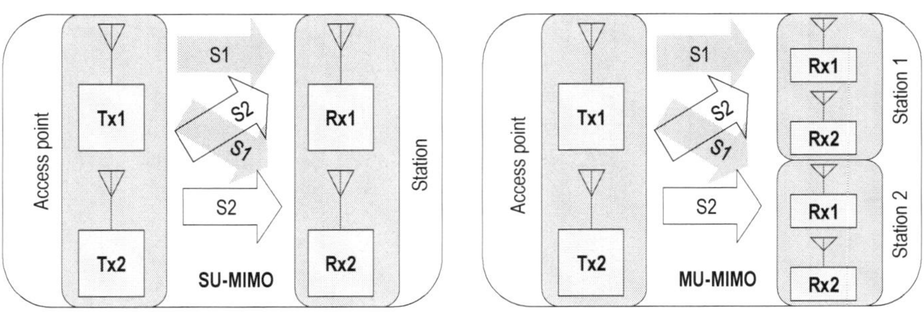

Figure 5.18. *SU-MIMO and MU-MIMO mechanism*

On the 802.11ac interface, the access point can also simultaneously transmit several spatially multiplexed streams to multiple stations using the MU-MIMO (Multi-User) mechanism (Figure 5.18).

5.3.5. *Modulation and coding scheme*

The value of the modulation and coding scheme (MCS) determines the rate value from the following parameters:

– modulation of the sub-carriers, phase modulation (BPSK or QPSK) or a mixed phase and amplitude modulation (16-QAM, 64-QAM or 256-QAM);

– coding rate of the error correcting code, which can take values of 1/2, 2/3, 3/4 or 5/6;

– number of spatial flows of the MIMO mechanism, which can take values between 1 and 8;

– bandwidth of the radio channel, which may be 20, 40, 80 or 160 MHz;

– duration of the guard interval, short duration of 400 ns or long duration of 800 ns.

Tables 5.4–5.7 respectively provide the rate values for a single spatial stream, for the MCS values between 0 and 9 and for bandwidths of 20, 40, 80 and 160 MHz.

Modulation	Coding rate	Number of bits per sub-carrier	Number of DATA bits per OFDM symbol	Number of encoded DATA bits	Rate (Mbps)	
					GI 800 ns	GI 400 ns
BPSK	1/2	1	26	52	6.5	7.2
QPSK	1/2	2	52	104	13.0	14.4
QPSK	3/4	2	78	104	19.5	21.7
16-QAM	1/2	4	104	208	26.0	28.9
16-QAM	3/4	4	156	208	39.0	43.3
64-QAM	2/3	6	208	312	52.0	57.8
64 QAM	3/4	6	234	312	58.5	65.0
64-QAM	5/6	6	260	312	65.0	72.2
256-QAM	3/4	8	312	416	78.0	86.7

Table 5.4. *Parameters of the modulation and coding scheme – Bandwidth of 20 MHz*

Modulation	Coding rate	Number of bits per sub-carrier	Number of DATA bits per OFDM symbol	Number of encoded DATA bits	Rate (Mbps)	
					GI 800 ns	GI 400 ns
BPSK	1/2	1	54	108	13.5	15.0
QPSK	1/2	2	108	216	27.0	30.0
QPSK	3/4	2	162	216	40.5	45.0
16-QAM	1/2	4	216	432	54.0	60.0
16-QAM	3/4	4	324	432	81.0	90.0
64-QAM	2/3	6	432	648	108.0	120.0
64 QAM	3/4	6	486	648	121.5	135.0
64-QAM	5/6	6	540	648	135.0	150.0
256-QAM	3/4	8	648	864	162.0	180.0
256-QAM	5/6	8	720	864	180.0	200.0

Table 5.5. *Parameters of the modulation and coding scheme – Bandwidth of 40 MHz*

Modulation	Coding rate	Number of bits per sub-carrier	Number of DATA bits per OFDM symbol	Number of encoded DATA bits	Rate (Mbps)	
					GI 800 ns	GI 400 ns
BPSK	1/2	1	117	234	29.3	32.5
QPSK	1/2	2	234	468	58.5	65.0
QPSK	3/4	2	351	468	87.8	97.5
16-QAM	1/2	4	468	936	117.0	130.0
16-QAM	3/4	4	702	936	175.5	195.0
64-QAM	2/3	6	936	1,404	234.0	260.0
64 QAM	3/4	6	1,053	1,404	263.3	292.5
64-QAM	5/6	6	1,170	1,404	292.5	325.0
256-QAM	3/4	8	1,404	1,872	351.0	390.0
256-QAM	5/6	8	1,560	1,872	390.0	433.3

Table 5.6. *Parameters of the modulation and coding scheme – Bandwidth of 80 MHz*

Modulation	Coding rate	Number of bits per sub-carrier	Number of DATA bits per OFDM symbol	Number of encoded DATA bits	Rate (Mbps)	
					GI 800 ns	GI 400 ns
BPSK	1/2	1	234	468	58.5	65.0
QPSK	1/2	2	468	936	117.0	130.0
QPSK	3/4	2	702	936	175.5	195.0
16-QAM	1/2	4	936	1,872	234.0	260.0
16-QAM	3/4	4	1,404	1,872	351.0	390.0
64-QAM	2/3	6	1,872	2,808	468.0	520.0
64 QAM	3/4	6	2,106	2,808	526.5	585.0
64-QAM	5/6	6	2,340	2,808	585.0	650.0
256-QAM	3/4	8	2,808	3,744	702.0	780.0
256-QAM	5/6	8	3,120	3,744	780.0	866.7

Table 5.7. *Parameters of the modulation and coding scheme Bandwidth of 160 MHz and 80+80 MHz*

6

Mutual Authentication

6.1. 802.1x mechanism

The 802.1x access control mechanism is deployed in the Local Area Network (LAN) implementing the following technologies:

– Ethernet technology in the case of access to a switch;

– Wireless Fidelity (Wi-Fi) in the case of a connection to an access point (AP).

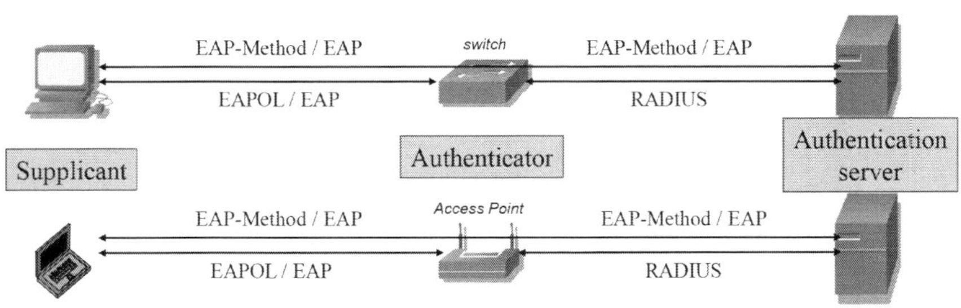

Figure 6.1. *Components of 802.1x mechanism*

The authentication uses the 802.1x access control mechanism that defines the following three components (Figure 6.1):

– the supplicant is the device (network host) wishing to access the Ethernet or Wi-Fi network;

– the authenticator is the device (Ethernet switch or Wi-Fi access point) that controls the supplicant's access to the LAN;

– the authentication server is the device that authenticates the supplicant and authorizes access to the LAN.

The 802.1x mechanism relies on the following set of protocols (Figure 6.2):

– the extensible authentication protocol (EAP) over LAN (EAPOL), exchanged between the supplicant and the authenticator;

– the EAP exchanged between the supplicant, on the one hand, and the authenticator or authentication server, on the other hand:

- the EAP is carried by the EAPOL protocol on the interface between the supplicant and the authenticator;

- the EAP carries EAP-Method messages exchanged between the supplicant and the authentication server;

– the remote authentication dial-in user service (RADIUS) protocol, exchanged between the authenticator and the authentication server. The RADIUS protocol carries the EAP on the interface between the authenticator and the authentication server.

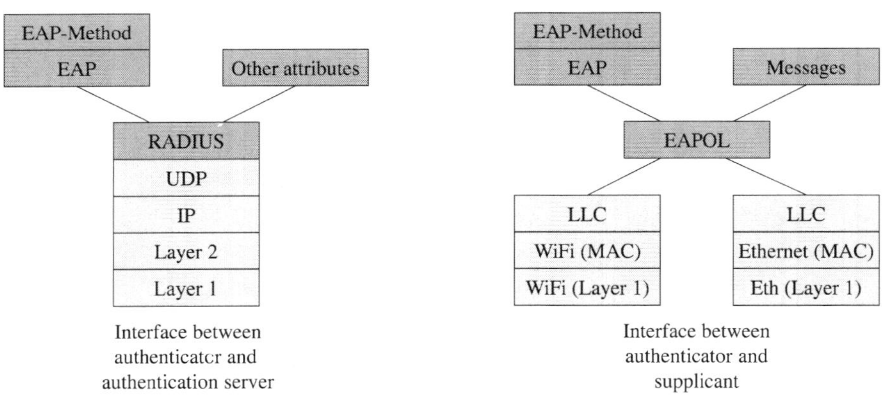

Figure 6.2. *Protocol architecture for 802.1x mechanism*

6.1.1. *EAPOL protocol*

The EAPOL protocol is exchanged between the supplicant and the authenticator. It initiates the supplicant's identity announcement and the capacities of each end. It ensures the transport of EAP/EAP-Method messages, which enable authentication of the supplicant, and possibly of the authentication server.

The structure of the EAPOL protocol is shown in Figure 6.3.

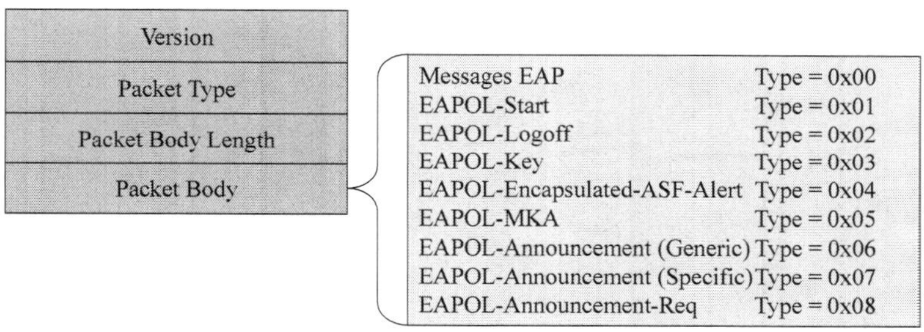

Figure 6.3. *Structure of EAPOL message*

The EAP is made up of a four-byte header and a packet body.

Version: this field, coded on one byte, identifies the version of the protocol and has the value of 03 in hexadecimal for the latest standardized version.

Packet Type: this field, coded one byte, identifies the type of data encapsulated by the EAPOL header. The EAPOL header can encapsulate an EAPOL message or an EAP message.

Packet Body Length: this field, coded on two bytes, indicates the size of the data encapsulated by the EAPOL header.

6.1.1.1. *EAPOL-Start message*

The EAPOL-Start message was transmitted without a message body in version 2 of the protocol. In version 3, the EAPOL-Start message can be transmitted with or without a message body. This message, transmitted by the supplicant, is used to initialize the 802.1x mechanism.

If the least significant bit of the first byte of the message body is set at ONE, the receiver of the EAPOL-Start message must make an announcement. The other bits of the first byte are set at ZERO.

The other bytes of the message body, if present, have a type, length, value (TLV) structure, which gives information about network access conditions.

6.1.1.2. *EAPOL-Logoff message*

The EAPOL-Logoff message is transmitted without a message body. This message, transmitted by the supplicant, is used to terminate the 802.1x mechanism. At the end of this message, the supplicant is no longer authenticated and its access to the LAN is blocked.

6.1.1.3. *EAPOL-Key message*

The EAPOL-Key message is transmitted by the supplicant or by the authenticator. It is used for the establishment of authentication and encryption keys derived from a master key.

The EAPOL-Key message is transmitted with a Key Descriptor message body containing the information necessary for key establishment.

6.1.1.4. *EAPOL-Encapsulated-ASF-Alert message*

The EAPOL-Encapsulated-ASF-Alert message is transmitted by the supplicant during authentication. It usually contains information specific to each constructor.

6.1.1.5. *EAPOL-Announcement message*

The EAPOL-Announcement message was introduced in version 3 for the transmission by the authenticator of information concerning network access conditions. The message body is composed of TLV structures.

The Network Identity TLV structure contains the identification of the network being subjected to access control. This structure can also be present in EAPOL-Start and EAPOL-Announcement-Req messages.

The Access Information TLV structure contains access information (access status, EAPOL messages processed). This structure can also be present in EAPOL-Start and EAPOL-Announcement-Req messages.

The Key Management Domain TLV structure contains the key-management domain name associated with a network name.

6.1.1.6. *EAPOL-Announcement-Req message*

The EAPOL-Announcement-Req message was introduced in version 3.

If the message body is absent or the least significant bit of the first byte of the message body is positioned at ONE, the receiver of the EAPOL-Announcement-Req message must make an announcement.

As for the EAPOL-Start message, the other bytes of the message body, if present, have a TLV structure that gives information on access conditions.

Unlike the EAPOL-Start message, the EAPOL-Announcement-Req message does not initiate the authentication procedure.

6.1.2. *EAP*

The EAP is deployed for the access of a supplicant (network host) to the authenticator (switch or access point) and the authentication server. It enables the transport of authentication data and does not require Internet Protocol (IP) connectivity.

The structure of the EAP is shown in Figure 6.4.

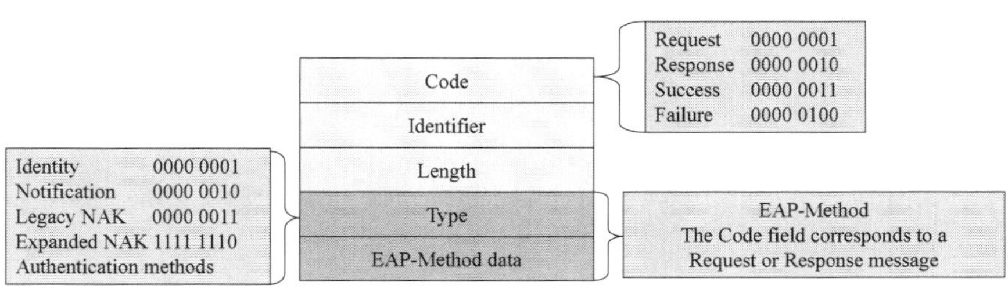

Figure 6.4. *EAP message structure*

The EAP is composed of a four-byte header and possibly an EAP-Method message.

Code: this field, coded on one byte, identifies the type of EAP message:

– Request: this message enables the authentication server to communicate with the supplicant, for example, to transmit to it an EAP-Method message; the supplicant can also transmit this message to the authentication server to request its identity during mutual authentication;

– Response: this message is sent in reply to the previous message; it can correspond, for example, to the supplicant's authentication data contained in an EAP-Method message;

– Success: this message is used by the authentication server to inform the supplicant that it has been successfully authenticated;

– Failure: this message is used by the authentication server to inform the supplicant that authentication has failed.

Identifier: this field, coded on one byte, enables correlation of the messages exchanged between the supplicant and the authentication server or the authenticator.

Length: this field, coded on two bytes, indicates the size of the EAP message. This value is identical to the value of the Packet Body Length field in the EAPOL protocol header.

When the EAP message is a request or a reply, the header encapsulates an EAP-Method message that contains the Type field, coded on one byte, identifying the type of data in the EAP-Method message.

The first three values in the Type field are reserved for specific messages (Identity, Notification and NAK). The other values pertain to identification methods such as EAP-message digest 5 (MD5) (Type = 4), EAP-TLS (Transport Layer Security) (Type = 13) or EAP-TTLS (Tunneled TLS) (Type = 21).

6.1.2.1. *EAP-Method Identity message*

The EAP-Method Identity message is used before or during the authentication phase. It is used to transport the supplicant's identity and, in some cases, during mutual authentication, the authenticator's identity. It can include data that will be presented to the user.

The EAP-Method Identity message is carried by the EAP Request message for the identity request and then by the EAP Response message for the reply containing the identity.

The EAP-Method Identity message is transferred to the authentication server. If the identity received is invalid, then this operation can be repeated several times. It is also possible for the authenticator to verify the supplicant's identity.

6.1.2.2. *EAP-Method Notification message*

The EAP-Method Notification message is used before or during the authentication phase. It is used to transport information on authentication status. It includes data that will be presented to the user.

The EAP-Method Notification message is carried by the EAP Request message for the notification request and then by the EAP Response message for the notification response.

6.1.2.3. *EAP-Method Legacy NAK message*

The EAP-Method Legacy NAK message is used by the supplicant to indicate that it does not support the authentication method suggested by the authentication server.

The EAP-Method Legacy NAK message is carried by the EAP Response message. It contains the authentication methods supported by the supplicant. If the message contains a Type field equal to ZERO, this means that the supplicant rejects the suggestion and that there is no alternative.

The EAP-Method Expanded NAK message is used by the supplicant in response to a request that also contains the EAP-Method Expanded NAK message. This functionality enables the number of types of authentication methods to be expanded beyond the 255 values allowed by the Type field.

6.1.3. *RADIUS messages*

The RADIUS protocol is used for transporting EAP-Method messages between the authentication server and the authenticator (switch or access point), used to identify the supplicant. RADIUS messages are encrypted and checked in their entirety using a secret shared between the two end points.

6.1.3.1. *Access-Request message*

The RADIUS Access-Request message is transmitted by the authenticator to the authentication server. This message is used to start the authentication procedure.

6.1.3.2. *Access-Challenge message*

The RADIUS Access-Challenge message is exchanged between the authenticator and the authentication server. This message is used to roll through the procedure for the authentication method.

6.1.3.3. *Access-Accept message*

The RADIUS Access-Accept message is one of the responses of the authentication server at the end of the authentication procedure. In this message, the authentication server accepts the supplicant's authentication request. Upon receipt of this message, the authenticator unblocks the supplicant's access.

6.1.3.4. *Access-Reject message*

The RADIUS Access-Reject message is one of the responses of the authentication server at the end of the authentication procedure. In this message, the authentication server rejects the supplicant's authentication request. Upon receipt of this message, the authenticator continues to block the supplicant's access.

6.1.4. *Authentication procedure*

Prior to the 802.1x authentication procedure, the supplicant must connect to the authenticator:

– if the supplicant is connected to an Ethernet switch, the 802.1x procedure starts when its interface is activated;

– if the supplicant is connected to a Wi-Fi access point, the 802.1x procedure starts at the end of the association phase with the Wi-Fi access point.

Whatever the authentication method used, the procedure is initiated by the supplicant, which transmits the EAPOL-Start message (Figure 6.5).

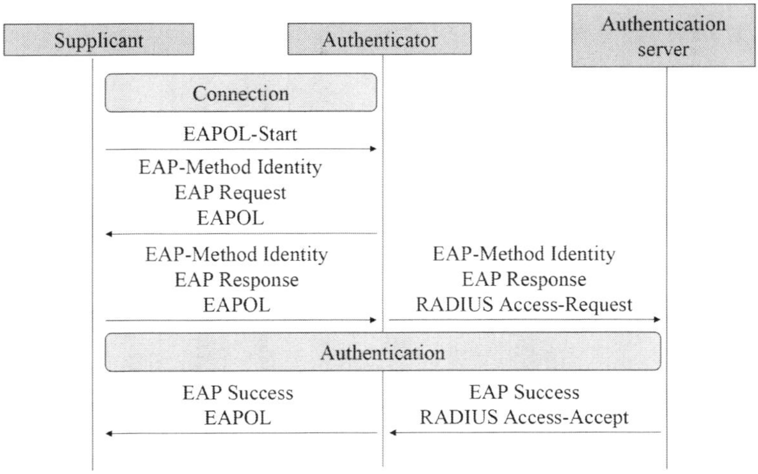

Figure 6.5. *Common exchanges in the authentication procedure*

The authenticator continues the procedure by sending the EAP-Request message containing the EAP-Method Identity message.

The supplicant provides its identity by replying with an EAP Response/EAP-Method Identity message. This message is transmitted by the authenticator to the authentication server in a RADIUS Access-Request message.

The next series of operations depend on the authentication method chosen.

After the authentication phase exchanges, the authentication server transmits to the supplicant:

– the EAP Success message if it is authenticated, in which case the authenticator authorizes traffic from the supplicant;

– the EAP Failure message in the opposite case, and access to the network remains prohibited.

The EAP Success (or EAP Failure) message is transmitted in a RADIUS Access-Accept (or Access-Reject) message at the interface between the authentication server and the authenticator.

6.2. Key management

6.2.1. *Key hierarchy*

Data protection on the radio interface is based mainly on secret keys. When a security association is established after successful authentication, temporary keys are created:

– pairwise transient key (PTK) is derived from the pairwise master key (PMK) for the unicast data;

– group transient key (GTK) is derived from the group master key (GMK) for the multicast and broadcast data.

These derived keys are regularly updated until the context is closed.

The derivation of the PMK uses the HMAC-SHA1 function, the result of which is 384 bits in size for the CCMP (Counter-mode/Cipher block chaining MAC (Message Authentication Code) Protocol) and 512 bits in size for the TKIP (Temporal Key Integrity Protocol).

The PTK, derived from the PMK, is obtained using the MAC addresses of the authenticator (AA) and the supplicant (SPA), and random numbers (Anonce and Snonce) exchanged during the four-way handshake procedure.

PTK = HMAC-SHA1(PMK, “Pairwise key expansion”, Min(AA,SPA) || Max(AA,SPA) || Min(ANonce,SNonce) || Max(ANonce,SNonce))

The PTK is cut up in order to provide the following keys:

– 128-bit key confirmation key (KCK). This key is used to authenticate messages during the four-way handshake procedure;

– 128-bit key encryption key (KEK). This key is used to encrypt messages during the four-way handshake and group key handshake procedures;

– 128-bit temporary key (TK). This key, used for TKIP and CCMP, serves to encrypt unicast data;

– 64-bit TMK1 (temporary message integrity code (MIC)) and TMK2. These keys, used for TKIP, check the integrity of the data. Each direction of transmission uses a specific key: TMK1 is used by the AP, and TMK2 is used by the station to generate the seal.

The derivation of the GMK also uses the HMAC-SHA1 function, with a result 128 bits in size for the CCMP and 256 bits in size for the TKIP.

The GTK, derived from the GMK master key, is obtained from the MAC address (AA) of the authenticator and from the random number (Gnonce):

GTK = HMAC-SHA1(GMK, "Group key expansion" || AA || GNonce)

The GTK is cut up in order to provide the following keys:

– 128-bit group encryption key (GEK). This key, used for TKIP and CCMP, encrypts broadcast and multicast data;

– 128-bit group integrity key (GIK). This key, used for TKIP, checks the integrity of broadcast and multicast data.

6.2.2. *Four-way handshake procedure*

The four-way handshake procedure defines four EAPOL-Key messages exchanged between the authenticator and the supplicant (Figure 6.6).

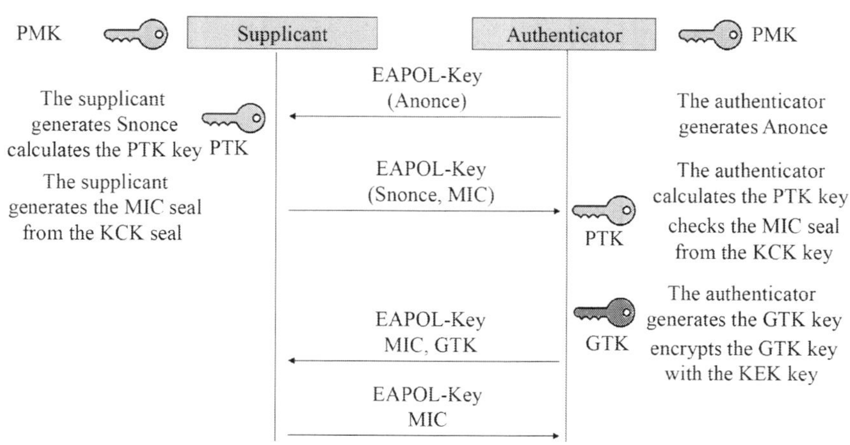

Figure 6.6. *Four-way handshake procedure*

This procedure enables the two end points to derive the PTK from the PMK and the distribution of the GTK by the authenticator.

The authenticator sends the first message to the supplicant if 802.1x authentication has been successful. This message contains the ANonce random number.

Upon reception of the first message, the supplicant generates a random number (SNonce), derives the PTK and constructs the second message containing the message integrity code (MIC) calculated from the KCK.

Upon reception of the second message, the authenticator derives the PTK, extracts the KCK from it and checks the MIC seal.

The third message is sent by the authenticator to the supplicant. It contains the GTK encrypted with the KEK and an MIC seal calculated using the KCK.

Upon reception of the third message, the supplicant checks that the MIC seal value is correct.

The fourth message is sent by the supplicant to complete the four-way handshake procedure.

Upon reception of the fourth message, the authenticator checks the MIC seal.

6.2.3. *Group Key Handshake procedure*

The group key handshake procedure (Figure 6.7) defines two EAPOL-Key messages exchanged between the authenticator and the supplicant. It takes place when the authenticator transmits a new GTK to the supplicant. The supplicant can start the procedure by sending an EAPOL-Key message.

The first message is initialized by the authenticator. It sends the new GTK encrypted with the KEK and the MIC seal calculated using the KCK.

Upon reception of the first message, the supplicant checks the MIC seal. It responds to the authenticator with the second message to acknowledge the first message.

Upon reception of the second message, the authenticator checks the MIC seal.

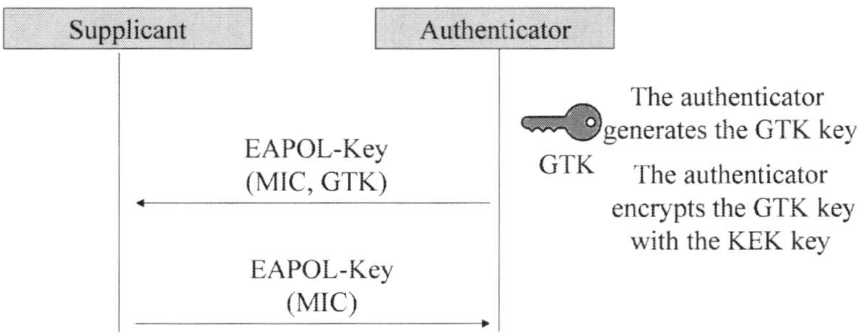

Figure 6.7. *Group key handshake procedure*

6.3. Application to the 4G mobile network

6.3.1. *EAP-AKA method*

The authentication and key agreement (AKA) mechanism has been defined for the attachment of the mobile to the 4G mobile network, and it allows mutual authentication of third parties and the distribution of keys.

Authentication is based on AUTN (Authentication Network) and RES (Result) seals generated by the home subscriber server (HSS) and the mobile from a RAND sequence and the secret key Ki.

The RAND sequence is generated by the HSS entity and then transmitted to the mobile. The secret key Ki is generated during the creation of the subscription and stored in the universal subscriber identity module (USIM) of the universal integrated circuit card (UICC) of the mobile.

Integrity key (IK) and cipher key (CK) are generated by the HSS entity and the mobile from a derivation of the Ki key using the RAND sequence. The master key PMK is derived from the keys CK and IK.

The EAP-AKA method is applied in the case of untrusted Wi-Fi access when establishing the SWu tunnel described in Chapter 7.

In the case of trusted Wi-Fi access, the EAP-AKA' method replaces the EAP-AKA method. The modification concerns the derivation of the keys CK and IK, which takes account of the identity of the access network, and the derivation algorithm.

The three components involved in the authentication procedure are integrated into the following entities:

– the supplicant is represented by the mobile which wishes to access the 4G mobile network;

– the authenticator is represented by the trusted Wi-Fi access that controls the access of the supplicant to the 4G mobile network;

– the authentication server is represented by the AAA (Authentication, Authorization and Accounting) server, which authenticates the supplicant and authorizes access to the 4G mobile network.

EAP-AKA' messages are carried between the trusted Wi-Fi access and the AAA server in DIAMETER messages:

– DER (Diameter-EAP-Request) message is transmitted by the trusted Wi-Fi access;

– DEA (Diameter-EAP-Answer) is transmitted by the AAA server.

6.3.2. *Mutual authentication procedure*

The procedure of mutual authentication, in the case of a trusted Wi-Fi access, is part of the procedure of attachment of the mobile.

At the end of the association phase with the trusted Wi-Fi access, the mobile transmits the EAPOL-Start message, which triggers the mutual authentication procedure based on the EAP-AKA' method (Figure 6.8).

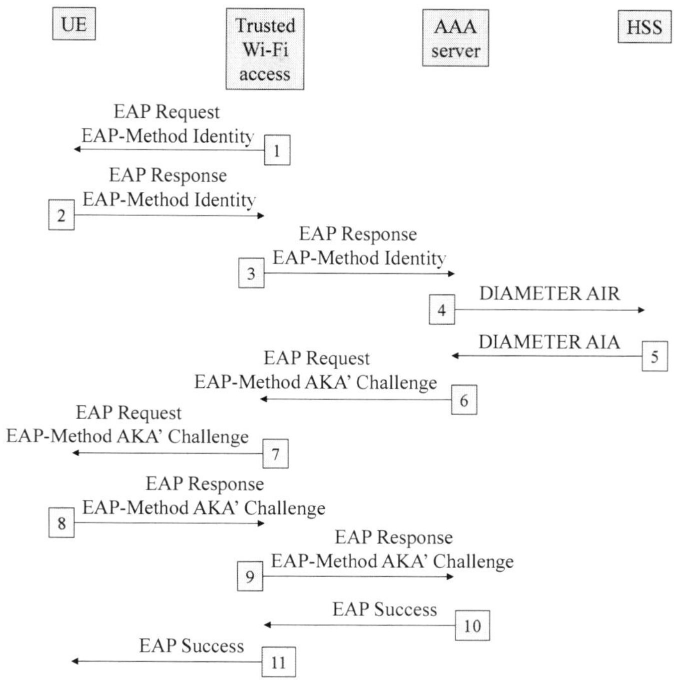

Figure 6.8. *Mutual authentication procedure*

1) Trusted Wi-Fi access sends the EAP Request message containing the EAP-Method Identity message.

2) The mobile transmits the EAP Response/EAP-Method Identity message containing, at the first authentication, the network access identifier (NAI) constructed from the international mobile subscriber identity (IMSI) of the mobile.

3) Wi-Fi access completes the EAP Response/EAP-Method Identity message, including the access network parameters (type, identity) and transfers it to the AAA server in a DER DIAMETER message.

4) The AAA server asks the HSS entity the cryptographic data of the mobile in the AIR (Authentication-Information-Request) DIAMETER message.

The HSS entity generates the RAND sequence and creates the RES, AUTN, CK' and IK' parameters from the key Ki and the RAND sequence.

5) The HSS entity transmits the authentication vectors to the AAA server in the AIA (Authentication-Information-Answer) DIAMETER message.

6) The AAA server derives the two keys CK' and IK' to generate the master key PMK and generates a pseudonym and possibly an identifier for the rapid renewal of the authentication.

The pseudonym and the identifier are temporary identities constructed from encryption of the private identity IMSI, using the advanced encryption standard (AES) algorithm. The same secret key is used by all AAA servers.

The AAA server transmits to the trusted Wi-Fi access the EAP Request/EAP-Method AKA' Challenge message containing the identity of the access network, the RAND sequence, the AUTN seal, the pseudonym and possibly the identifier for the renewal of the authentication.

This message is transmitted in a DEA DIAMETER message and contains a message authentication code (MAC) for the integrity check.

7) Trusted Wi-Fi access transfers the EAP Response/EAP-Method AKA' Challenge message to the mobile.

8) The mobile locally calculates, from its key Ki and the received RAND number, the key PMK, its seal RES and that of the AUTN network. The mobile compares the received AUTN with the calculated value. If both values are the same, the network is authenticated. The mobile also controls the integrity of the received message.

The mobile transmits the EAP Response/EAP-Method AKA' Challenge message containing the RES seal and the MAC seal for the integrity check of the message to the trusted Wi-Fi access.

9) Trusted Wi-Fi access transfers the EAP Response/EAP-Method AKA' Challenge message to the AAA server in a DER DIAMETER message.

10) The AAA server checks the integrity of the received message and compares the RES seal received from the mobile to that received from the HSS entity. If the two values are identical, the mobile is authenticated.

The AAA server transmits the DEA DIAMETER message containing the EAP Success message and the PMK to the trusted Wi-Fi access.

11) Trusted Wi-Fi access stores the PMK and transfers the EAP Success message to the mobile.

6.3.3. *Procedure for rapid renewal of authentication*

The rapid renewal of authentication makes it possible to avoid repeating the procedure from the authentication vector (RAND, AUTN, RES, CK', IK').

The implementation of the procedure for rapid renewal of authentication is indicated by the AAA server, during the initial authentication procedure, when it supplies the corresponding identifier.

The identity of the access network must not change during the procedure for rapid renewal of authentication. If this happens, the normal authentication procedure must be carried out.

The procedure for rapid renewal of the authentication is described in Figure 6.9.

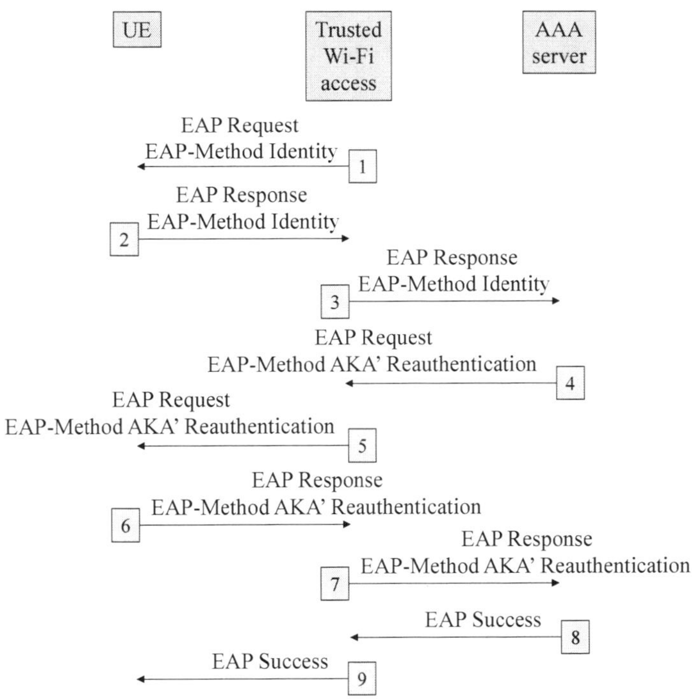

Figure 6.9. *Procedure for rapid renewal of authentication*

Steps 1 to 3 are identical to those described for initial authentication in Figure 6.8. The private identity used by the mobile is the identifier for rapid renewal of authentication.

4) The AAA server transmits to the trusted Wi-Fi access the EAP Request/EAP-Method AKA' Reauthentication message containing a random number NOUNCE for the generation of a new PMK and a new identifier for the next authentication.

This message is transmitted in a DER DIAMETER message and contains a MAC seal for the integrity check.

5) Trusted Wi-Fi access transfers the EAP Response/EAP-Method AKA' Reauthentication message to the mobile.

6) The mobile checks the integrity of the received message and acknowledges it in the EAP Response/EAP-Method AKA' Reauthentication message containing a MAC seal.

7) Trusted Wi-Fi access transfers the EAP Response/EAP-Method AKA' Reauthentication message to the AAA server in a DER DIAMETER message.

Steps 8 and 9 are identical to those described for initial authentication in Figure 6.8.

6.3.4. *Application to the MIPv4 FA mechanism*

The MIPv4 FA (Mobile IP version 4 Foreign Agent) mechanism is an alternative for building the S2a tunnel containing the mobile stream.

The MIPv4 FA mechanism defines the following three components:

– the mobile node (MN) component integrated in the mobile;

– the home agent (HA) component integrated in the PDN Gateway (PGW);

– the foreign agent (FA) component integrated into an entity (e.g. a router) of the Wi-Fi access network, which is not necessarily the trusted Wi-Fi access.

During the mutual authentication procedure, the AAA server and the mobile also generate the extended master session key (EMSK) from the two keys CK' and IK'.

Two keys, MN-HA and MN-FA, are generated from the EMSK to protect the MIPv4 messages exchanged between, on the one hand, the component MN and, on the other hand, the components HA and FA.

1) The AAA server and the mobile derive the EMSK to generate the MIP-RK.

2) The AAA server and the mobile derive the MIP-RK to generate the FA-RK. The AAA server transfers the FA-RK to the trusted Wi-Fi access.

3) The AAA server and the mobile derive the key MIP-RK to generate the key MN-HA. The AAA server transfers the MN-HA key to the PGW entity.

4) The mobile and the trusted Wi-Fi access derive the FA-RK to generate the MN-FA key. Trusted Wi-Fi access transfers the MN-FA key to the FA component.

7

SWu Tunnel Establishment

7.1. IPSec mechanism

The IPSec (Internet Protocol Security) mechanism offers security services (authentication, integrity and confidentiality) in an identical way in IPv4 and IPv6. Their implementation is optional in IPv4 but mandatory in IPv6. Their use is optional.

Security services are offered through the use of AH (Authentication Header) or ESP (Encapsulating Security Payload) extensions of the IPv4 or IPv6 header.

To secure a two-directional communication between two end points, a security association (SA) pair is required. The IKEv2 (Internet Key Exchange version 2) protocol dynamically ensures the creation of the security association.

A security association contains the following parameters:

– the authentication algorithm and the key in order to generate the AH extension;

– the encryption algorithm and the key in order to generate the ESP extension;

– the authentication algorithm and the key in order to generate the ESP extension, if this service is used;

– the lifetime of the security association;

– the encapsulation mode (tunnel or transport).

The IPSec mechanism defines the following three databases:

– security policy database (SPD): this defines the security policy to be applied to input and output traffic for a host or a security gateway;

– security association database (SAD): this contains the parameters applied to a security association;

– peer authorization database (PAD): this provides a link between the IKEv2 protocol and the SPD.

The selector is the mechanism enabling identification at the source of the security association to be applied to traffic. The selector uses the following fields:

– Protocol (IPv4) or Next Header (IPv6) and source or destination address of the IP headers;

– source or destination port of the TCP (Transmission Control Protocol) or UDP (User Datagram Protocol) headers.

For an outgoing packet, the selector consults the SPD that defines the process to be applied to the packet:

– BYPASS: the packet is transmitted without a security service;

– DISCARD: the packet is discarded;

– PROTECT: the security service is applied to the packet. If the security association is not established, then the IKE protocol is invoked. If the security association exists, then the database returns a pointer to the SAD.

The deletion of a packet causes the generation from the security gateway toward the source of an ICMP message with the following characteristics:

– in the IPv4 environment, Type = 3 (destination unreachable) and Code = 13 (Communication Administratively Prohibited);

– in the IPv6 environment, Type = 1 (destination unreachable) and Code = 1 (Communication with Destination Administratively Prohibited).

The security parameter index (SPI) is a field in the AH or ESP header used by the destination to identify the security association in a unique way. The destination uses this index to extract the security association parameters from the SAD.

For an incoming packet, the SPD is consulted if the packet is not protected, and the instruction (BYPASS or DISCARD) is applied. If the IP packet is protected, then the SPI field is used to recover the parameter of the security association.

7.1.1. *Header extensions*

The IPSec mechanism introduces two IPv4 or IPv6 header extensions:

– authentication header (AH) is designed to ensure the integrity and authentication of IP packets without data encryption (no confidentiality);

– encapsulating security payload (ESP) ensures the integrity, authentication and confidentiality of IP packets.

7.1.1.1. *AH extension*

The presence of the AH extension is indicated by the Next Header (in IPv6) or Protocol (in IPv4) field of the previous header, with a value of 51.

The AH extension contains the following fields (Figure 7.1).

```
0                   1                   2                   3
0 1 2 3 4 5 6 7 8 9 0 1 2 3 4 5 6 7 8 9 0 1 2 3 4 5 6 7 8 9 0 1
+-+-+-+-+-+-+-+-+-+-+-+-+-+-+-+-+-+-+-+-+-+-+-+-+-+-+-+-+-+-+-+-+
| Next Header |  Payload Len  |             RESERVED            |
+-+-+-+-+-+-+-+-+-+-+-+-+-+-+-+-+-+-+-+-+-+-+-+-+-+-+-+-+-+-+-+-+
|                Security Parameters Index (SPI)                |
+-+-+-+-+-+-+-+-+-+-+-+-+-+-+-+-+-+-+-+-+-+-+-+-+-+-+-+-+-+-+-+-+
|                    Sequence Number Field                      |
+-+-+-+-+-+-+-+-+-+-+-+-+-+-+-+-+-+-+-+-+-+-+-+-+-+-+-+-+-+-+-+-+
|                                                               |
+               Integrity Check Value-ICV (variable)            |
|                                                               |
+-+-+-+-+-+-+-+-+-+-+-+-+-+-+-+-+-+-+-+-+-+-+-+-+-+-+-+-+-+-+-+-+
```

Figure 7.1. *AH extension format*

Next Header: this field, coded on one byte, indicates the type of header following the ESP extension.

Payload Length: this field, coded on 1 byte, provides the extension size in multiples of four bytes, not including the first eight bytes. The size of the extension in IPv6 must remain a multiple of eight bytes.

Security parameters index (SPI): this field, coded on four bytes, contains a value pertaining to the previously negotiated security association.

Sequence Number: this field, coded on four bytes, contains a value increased by one unit for each IPv4 or IPv6 packet transmitted. This field enables protection against replay. This field has a value of 1 for the first packet transmitted. When the counter reaches the maximum value, a new security association must be negotiated in order to avoid the start of a new cycle.

An extended sequence number (ESN), coded on eight bytes, constitutes an option, making it possible for the lifetime of the security association to be prolonged. In order to preserve the structure of the extension, the 32 least significant bits are transmitted in the sequence number field. However, the seal is calculated on all 64 bits.

Integrity check value (ICV): this field is coded on a multiple of four bytes and contains the seal of the data, ensuring authentication and integrity checking.

7.1.1.2. *ESP extension*

The presence of the ESP extension (Figure 7.2) is indicated by the Next Header (in IPv6) or Protocol (in IPv4) field of the previous header, with a value of 50.

```
0                   1                   2                   3
0 1 2 3 4 5 6 7 8 9 0 1 2 3 4 5 6 7 8 9 0 1 2 3 4 5 6 7 8 9 0 1
+-+-+-+-+-+-+-+-+-+-+-+-+-+-+-+-+-+-+-+-+-+-+-+-+-+-+-+-+-+-+-+-+
|               Security Parameters Index (SPI)                 |
+-+-+-+-+-+-+-+-+-+-+-+-+-+-+-+-+-+-+-+-+-+-+-+-+-+-+-+-+-+-+-+-+
|                      Sequence Number                          |
+-+-+-+-+-+-+-+-+-+-+-+-+-+-+-+-+-+-+-+-+-+-+-+-+-+-+-+-+-+-+-+-+
|                    Payload Data (variable)                    |
~                                                               ~
|                                                               |
+               +-+-+-+-+-+-+-+-+-+-+-+-+-+-+-+-+-+-+-+-+-+-+-+-+
|               |             Padding (0-255 bytes)             |
+-+-+-+-+-+-+-+-+               +-+-+-+-+-+-+-+-+-+-+-+-+-+-+-+-+
|                               |  Pad Length   |  Next Header  |
+-+-+-+-+-+-+-+-+-+-+-+-+-+-+-+-+-+-+-+-+-+-+-+-+-+-+-+-+-+-+-+-+
|          Integrity Check Value-ICV (variable)                 |
~                                                               ~
|                                                               |
+-+-+-+-+-+-+-+-+-+-+-+-+-+-+-+-+-+-+-+-+-+-+-+-+-+-+-+-+-+-+-+-+
```

Figure 7.2. *ESP extension format*

The ESP extension contains the same fields as the AH extension. It starts with the SPI and Sequence Number fields. After these fields come the encapsulated data, which may contain synchronization data (initialization vector) of the encryptor. Following the encapsulated data, the extension ends with the following fields: Padding, Pad, Length, Next Header and optionally ICV (authentication, data).

The Padding field is necessary when block encryption is used, and the block must be of a certain size, to align the packet size with a multiple of four bytes.

7.1.1.3. *Transport and tunnel modes*

For transport mode, the AH or ESP header is inserted between the IP header and the source IP packet payload. In the IPv6 environment, the AH or ESP header appears after the Hop-by-Hop, Destination, Routing and Fragment extensions.

For tunnel mode, the AH or ESP header encapsulates the source IP packet, and the whole is encapsulated in its turn by a new IP header. The tunnel corresponds to a data structure, in which an IP packet contains another IP packet.

When the AH header is used, authentication is applied to the whole packet except for the variable fields of the IP header (Figure 7.3).

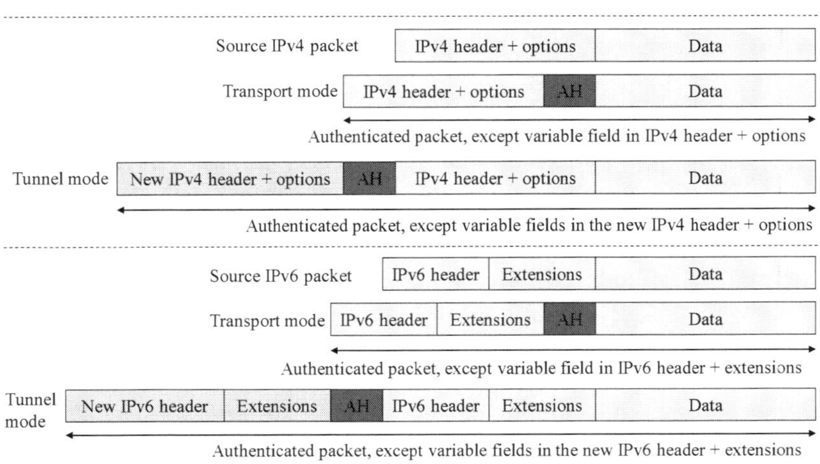

Figure 7.3. *Position of AH extension*

The variable fields of the IPv4 header are set at ZERO to calculate the authentication digest:

– DSCP (DiffServ Code Point): the value of this field can be modified by an intermediary router when it checks traffic characteristics;

– ECN (Explicit Congestion Notification): the value of this field can be modified by an intermediary router to alert the destination that congestion is developing;

– DF (Don't Fragment): this bit can be set at ONE by an intermediary router;

– Fragment Offset: insertion of the AH header occurs on non-fragmented IP packets, and, therefore, this field has a value of zero;

– TTL (Time To Live): the value of this field is decreased by one unit for each router crossed;

– Checksum: the value of this field is recalculated as soon as a field in the IP header changes value.

The set of IPv4 header options is considered as a single entity. Some options can be modified by an intermediary router. If a single modifiable option appears, then the set of options is set at ZERO for the authentication digest calculation.

The variable fields of the IPv6 header are set at ZERO for the authentication digest calculation. These are identical fields to the ones in the IPv4 header (DSCP, ECN and Hop Limit), as well as the Flow Label field.

The Hop-by-Hop and Destination extensions of the IPv6 header have a bit that indicates whether the option can be modified by an intermediary router or not. If this bit is set at ONE, then the extension is set at ZERO for the authentication digest calculation.

When the ESP header is used in transport mode, the confidentiality service is applied to the encapsulated data and ESP tail. Authentication and integrity services cover the ESP header, encapsulated data and ESP tail (Figure 7.4).

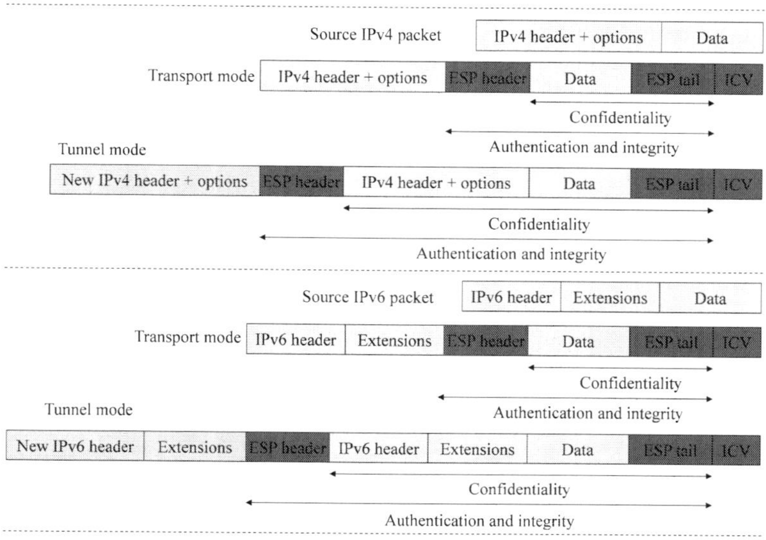

Figure 7.4. *Position of ESP extension*

When the ESP header is used in tunnel mode, the confidentiality service is applied to the source IP packet and ESP tail. Authentication and integrity services cover the ESP header, source IP packet and ESP tail (Figure 7.4).

Note that in transport mode, the Destination extension can appear before, after or simultaneously before and after the AH or ESP extension.

7.1.2. *IKEv2 protocol*

The IKEv2 protocol is more simplified than the previous version. It combines the functionalities defined in IKEv1 and Internet security association and key management protocol (ISAKMP) while removing unnecessary processes. It eliminates the generic character of the previous version, integrating the domain of interpretation (DOI) function, which defines the parameters specific to the ESP/AH security association.

Each IKEv2 message is composed of a header (HDR) and a sequence of blocks. The IKEv2 message is encapsulated by a UDP header with source and destination port values of 500 or 4500. When the 4500 port is used, the IKEv2 message is preceded by four bytes at ZERO.

7.1.2.1. *Message header*

The header of the IKE message contains the following fields (Figure 7.5):

```
                     1                   2                   3
 0 1 2 3 4 5 6 7 8 9 0 1 2 3 4 5 6 7 8 9 0 1 2 3 4 5 6 7 8 9 0 1
+-+-+-+-+-+-+-+-+-+-+-+-+-+-+-+-+-+-+-+-+-+-+-+-+-+-+-+-+-+-+-+-+
|                    IKE SA Initiator's SPI                     |
|                                                               |
+-+-+-+-+-+-+-+-+-+-+-+-+-+-+-+-+-+-+-+-+-+-+-+-+-+-+-+-+-+-+-+-+
|                    IKE SA Responder's SPI                     |
|                                                               |
+-+-+-+-+-+-+-+-+-+-+-+-+-+-+-+-+-+-+-+-+-+-+-+-+-+-+-+-+-+-+-+-+
| Next Payload  | MjVer | MnVer | Exchange Type |     Flags     |
+-+-+-+-+-+-+-+-+-+-+-+-+-+-+-+-+-+-+-+-+-+-+-+-+-+-+-+-+-+-+-+-+
|                          Message ID                           |
+-+-+-+-+-+-+-+-+-+-+-+-+-+-+-+-+-+-+-+-+-+-+-+-+-+-+-+-+-+-+-+-+
|                            Length                             |
+-+-+-+-+-+-+-+-+-+-+-+-+-+-+-+-+-+-+-+-+-+-+-+-+-+-+-+-+-+-+-+-+
```

Figure 7.5. *IKE message header format*

Initiator's SPI: this field, coded on eight bytes, incorporates a value chosen by the initiator. It initializes identification of the IKE security association.

Responder's SPI: this field, coded on eight bytes, incorporates a value chosen by the responder. It completes the identification of the IKE security association.

Next Payload: this field, coded on one byte, incorporates the indication of the type of block following the header (Table 7.1).

Major Version: this field, coded on four bits, indicates the maximum value of the IKE protocol version that can be used. This value is equal to 2 for the implementation of the IKEv2 protocol.

Minor Version: this field, coded on four bits, indicates the minimum value of the IKE protocol version. This value is equal to 0 for the implementation of the IKEv2 protocol.

Exchange Type: this field, coded on one byte, indicates the type of exchange to which the message belongs:

– IKE_SA_INIT: this exchange concerns the first phase of the establishment of the IKE security association;

– IKE_AUTH: this exchange concerns the second phase of the establishment of the security association;

– CREATE_CHILD_SA: this exchange concerns the establishment of the ESP/AH security association;

– INFORMATIONAL: this exchange concerns event notification.

Each type of exchange imposes a certain number of required blocks composing the message and defines optional blocks.

Notation	Designation
SA	Security Association
KE	Key Exchange
Idi	Identification – initiator
IDr	Identification – responder
CERT	Certificate
CERTREQ	Certificate Request
AUTH	Authentication
Ni	Nonce – initiator
Nr	Nonce – responder
N	Notification
D	Delete
V	Vendor ID
TSi	Traffic Selector – initiator
TSr	Traffic Selector – responder
SK	Encrypted and Authenticated
CP	Configuration
EAP	Extensible Authentication

Table 7.1. *Block types*

Flags: this field includes the following three flags:

– R (Response): this flag, positioned at ONE, indicates that this message is a response. An IKE termination must not respond to a response except when authentication has failed;

– V (Version): this flag, positioned at ONE, indicates that the IKE termination is able to process a version higher than the one shown in the Major Version field;

– I (initiator): this flag, positioned at ONE, indicates that the message is generated by the initiator of the IKE security association.

Message ID: this field, coded on four bytes, is an identifier used to control the retransmission of lost messages and to correlate the request and response. It also protects against replay attacks.

Length: this field, coded on four bytes, includes the IKE message size.

7.1.2.2. *Blocks*

Each block starts with a generic header containing the Next Payload field, the C (Critical) bit and the Payload Length field (Figure 7.6).

```
 0                   1                   2                   3
 0 1 2 3 4 5 6 7 8 9 0 1 2 3 4 5 6 7 8 9 0 1 2 3 4 5 6 7 8 9 0 1
+-+-+-+-+-+-+-+-+-+-+-+-+-+-+-+-+-+-+-+-+-+-+-+-+-+-+-+-+-+-+-+-+
| Next Payload  |C|  Reserved   |         Payload Length        |
+-+-+-+-+-+-+-+-+-+-+-+-+-+-+-+-+-+-+-+-+-+-+-+-+-+-+-+-+-+-+-+-+
```

Figure 7.6. *Format of generic block header*

The Next Payload field indicates the type of block that comes next, thus enabling chaining.

The C bit determines the process to be executed when the receiver does not recognize the block:

– if the C bit is positioned at ONE, the receiver rejects the message;

– if the C bit is positioned at ZERO, the receiver ignores the block and processes the next block.

7.1.2.2.1. SA block

The SA block is used for the negotiation of the parameters of IKE and ESP/AH security association. An SA block can contain several proposals (P) ranked in order of preference. Each proposal defines a protocol (IKE, ESP or AH) and the value of the SPI. Each proposal includes several transformations (T), and each transformation includes one or more attributes (A).

The transformation T involves the following operations:

– the encryption algorithm ENCR: this operation is used for negotiation pertaining to the IKE and ESP;

– the pseudo-random function PRF: this operation is used for negotiation pertaining to the IKE;

– the integrity algorithm INTEG: this operation is used for negotiation pertaining to IKE, AH and (optionally) ESP;

– the Diffie–Hellman (D-H) group: this operation is used for negotiation pertaining to IKE and (optionally) AH and ESP;

– the extended sequence number ESN: this operation is used for negotiation pertaining to the AH and ESP.

The attribute A specifies the length of the encryption algorithm key defined in the ENCR transformation. The other transformations, PRF, INTEG, D-H and ESN have no attributes.

7.1.2.2.2. KE block

The key exchange (KE) block contains the public Diffie–Hellman value, enabling each end point (initiator and responder) to construct a shared secret. The block also mentions the D-H group defined in the SA block.

7.1.2.2.3. IDi and IDr blocks

Identification initiator (IDi) and identification responder (IDr) blocks contain an identification of the initiator of the IKE message and the responder. This identification is based on an IPv4 or IPv6 address, a name, a messaging address or a group of bytes.

7.1.2.2.4. CERT block

The certificate (CERT) block provides a means of transporting a certificate or information pertaining to authentication.

7.1.2.2.5. CERTREQ block

The certificate request (CERTREQ) block is a request pertaining to a certificate. It is used in the response of the IKE_INIT_SA exchange response or in the IKE_AUTH exchange request. It also indicates the certification authority for the required certificate.

7.1.2.2.6. AUTH block

The authentication (AUTH) block contains the authentication digest (signature or seal) for the third-party authentication. The block also specifies the method used.

7.1.2.2.7. Ni and Nr blocks

Nonce initiator (Ni) and nonce responder (Nr) blocks contain a random number generated by the initiator and responder. These numbers are used in the creation of derived keys.

7.1.2.2.8. N block

The notification (N) blocks contain error messages indicating the reason why the security association cannot be established.

The N block also contains status messages that an SA management process wishes to communicate to a remote process.

7.1.2.2.9. D block

The delete (D) block includes the SPI of the SA that the message source wishes to delete. For an AH or ESP, it is possible to specify several SPI values. It is also possible to string several D blocks together in a single IKEv2 message.

7.1.2.2.10. V block

The vendor ID (V) block announces that the message source is capable of accepting private extensions of the IKEv2 protocol. These extensions can involve the introduction of new blocks, new types of exchange or new notification information.

7.1.2.2.11. TS block

The traffic selector (TS) block identifies the flows for which the ESP/AH security association is implemented. Flow determination is based on the following information:

– the type of data encapsulated by the IP header, stated in the Protocol fields of the IPv4 header or the Next Header field of the IPv6 header;

– the range of source and destination IP addresses;

– the range of source and destination port numbers if the IP header encapsulates UDP or TCP segments;

– the ICMP message type and code.

7.1.2.2.12. SK block

The SK (encrypted and authenticated) block is always located at the end of the IKEv2 message. The encryption and integrity algorithms for the IKEv2 message are negotiated during the implementation of the IKEv2 SA.

7.1.2.2.13. CP block

The configuration (CP) block is used to exchange configuration information between the two end points. In the case of an ESP/AH security association between a host and a security gateway, the host can request information concerning a host in the protected network.

7.1.2.2.14. EAP block

The extensible authentication protocol (EAP) block enables the use of EAP in IKEv2 message for authentication.

7.1.3. *Procedure*

7.1.3.1. *IKE_SA_INIT exchange*

The first exchange, IKE_SA_INIT, negotiates cryptographic algorithms and random numbers and executes a Diffie–Hellman exchange in order to create an IKE security association.

The initiator generates an IKE message containing the header (HDR) and the SAi1, KEi and Ni blocks. The header contains the initiator's SPI, version numbers and flags. The SAi1 block contains the cryptographic algorithms proposed by the initiator for the IKE security association. The KEi block includes the Diffie–Hellman group and public value. The Ni block displays the initiator's random number (Figure 7.7).

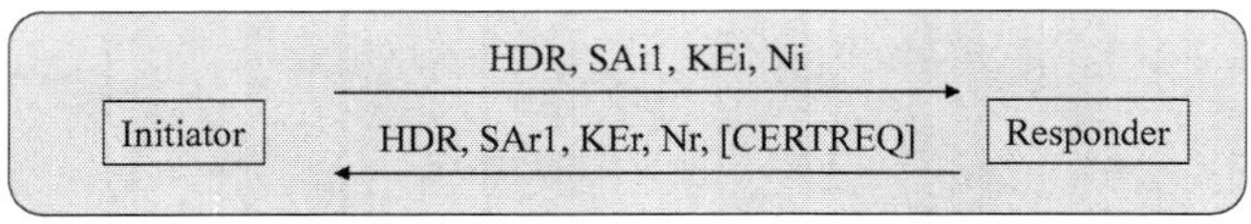

Figure 7.7. *IKE_SA_INIT exchange*

The responder chooses a cryptograph series from the initiator's proposals and includes it in the SAr1 block. It completes the exchange of Diffie–Hellman keys with the KEr block. It sends its random number in the Nr block (Figure 7.7). It can possibly communicate a list of certificate authorities in the CERTREQ block.

At this stage of the negotiation, each end point can generate the SKEYSEED key. The keys used for encryption and integrity of IKE messages are produced by the SKEYSEED key and are known as SK_e (encryption) and SK_a (integrity). Message protection involves only the blocks; the header is not included.

The two different directions of traffic use different keys. The keys used to protect messages from the initiator are SK_ai and SK_ei. The keys used to protect messages in the other direction are SK_ar and SK_er.

Other keys are also derived from the SKEYSEED key. The SK_d key is used to derive the keys used in the ESP/AH security association.

The SKEYSEED key is calculated using the Diffie–Hellman secret (D-H key) and the random numbers Ni and Nr:

SKEYSEED = PRF (D-H key, Ni | Nr)

The keys SK_d, SK_ai, SK_ar, SK_ei, SK_er, SK_pi and SK_pr are generated as follows:

SK_d = PRF (SKEYSEED, Ni | Nr | SPIi | SPIr | 0x01);

SK_ai = PRF (SKEYSEED, SK_d | Ni | Nr | SPIi | SPIr | 0x02);

SK_ar = PRF (SKEYSEED, SK_ai | Ni | Nr | SPIi | SPIr | 0x03);

SK_ei = PRF (SKEYSEED, SK_ar | Ni | Nr | SPIi | SPIr | 0x04);

SK_er = PRF (SKEYSEED, SK_ei | Ni | Nr | SPIi | SPIr | 0x05).

7.1.3.2. *IKE_AUTH exchange*

The second exchange, IKE_AUTH, is used to authenticate previous IKE messages and communicate identities and possibly exchange certificates, as well as to establish the first AH/ESP security association. These messages are completely encrypted and protected by the keys established during the IKE_SA_INIT exchange.

The initiator indicates its identity with the IDi block. It authenticates its identity and protects the integrity of the first message in the IKE_SA_INIT exchange using the AUTH block.

It can possibly send its certificate in the CERT block and the certification authority's identity in the CERTREQ block. In this case, the AUTH block contains the initiator's signature (Figure 7.8).

Figure 7.8. *IKE_AUTH exchange*

The optional IDr block enables the initiator to specify which of the responder's identities it wishes to communicate with (Figure 7.8).

The initiator starts the negotiation of the AH/ESP security association with the SAi2 block. The TSi block specifies the characteristics of the packets transferred by the initiator. The TSr block specifies the address for packets transferred to the responder (Figure 7.8).

The notation SK{ ... } indicates that the blocks are completely encrypted and protected (Figure 7.8).

The responder communicates its identity in the IDr block. It may send a certificate. It authenticates its identity and protects the integrity of the second message of the IKE_SA_INIT exchange. It completes the negotiation of the ESP/AH security association with the SAr2 block (Figure 7.8).

7.1.3.3. *CREATE_CHILD_SA exchange*

The CREATE_CHILD_SA exchange is used to create the ESP/AH security association and to renew the keys of the IKE and ESP/AH security association.

For the exchange involving the creation of the ESP/AH security association, the initiator finalizes it in the SA block and the traffic selectors proposed for the SA in the TSi and TSr blocks. It transmits a random number in the Ni block and optionally a Diffie–Hellman value in the KEi block (Figure 7.9).

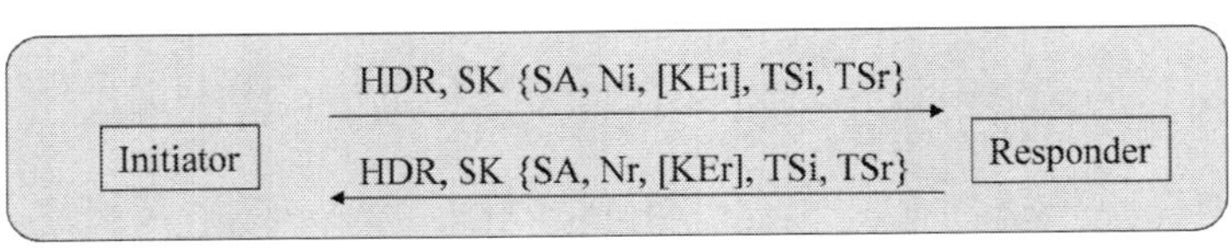

Figure 7.9. *CREATE_CHILD_SA exchange creation of ESP/AH SA*

The responder confirms the offer in the SA block. It transmits a Diffie–Hellman value in the KEr block, if the KEi block has been included in the request (Figure 7.9).

The KEYMAT key, used for the ESP/AH security association, is derived from the SK_d key and the random numbers Ni and Nr. If the exchange contains Diffie–Hellman values, then the secret D-H key obtained also participates in the creation of the KEYMAT key:

KEYMAT = PRF (SK_d, Ni | Nr)

KEYMAT = PRF (SK_d, D-H key | Ni | Nr)

To renew the IKE SA key, the initiator sends the parameters in the SA block, a random number in the Ni block and a Diffie–Hellman value in the KEi block. The initiator's new SPI is provided in the SA block (Figure 7.10).

The responder confirms the offer in the SA block. It transmits a Diffie–Hellman value in the KEr block. The responder's new SPI is provided in the SA block (Figure 7.10).

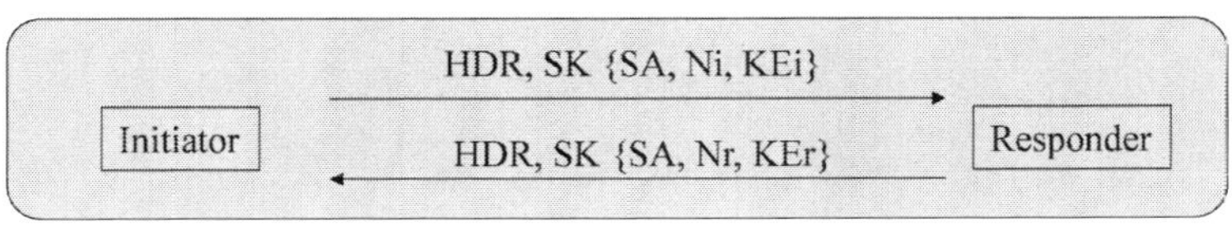

Figure 7.10. *CREATE_CHILD_SA exchange renewal of IKE SA key*

The new SKEYSEED key is computed from the old key SK_d, the secret key D-H and the random numbers Ni and Nr.

SKEYSEED = PRF (old SK_d, D-H key | Ni | Nr)

To renew the ESP/AH SA key, the messages transmitted by the initiator and the responder are similar to the ones used in the creation of the security association. The initiator's request contains an N block (REKEY_SA) containing the SPI value of the new security association (Figure 7.11).

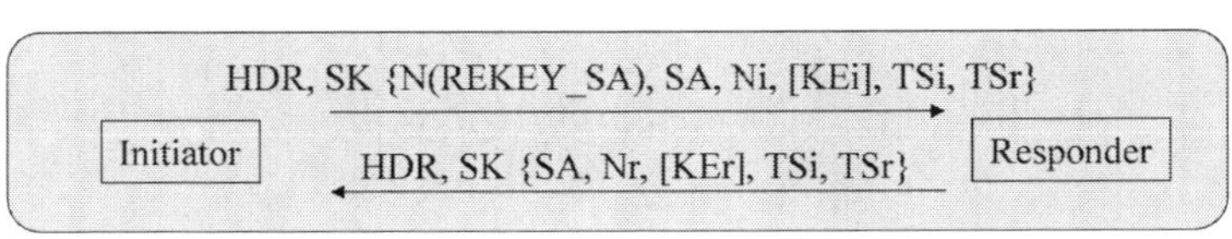

Figure 7.11. *CREATE_CHILD_SA exchange renewal of ESP/AH SA key*

7.2. Application to the 4G mobile network

7.2.1. *SWu tunnel establishment procedure*

The IPSec mechanism is implemented for the establishment of the SWu tunnel, at the end of the authentication phase using the 802.1x mechanism described in Chapter 6, the mobile acting as the initiator and the evolved packet data gateway (ePDG) of the responder (Figure 7.12).

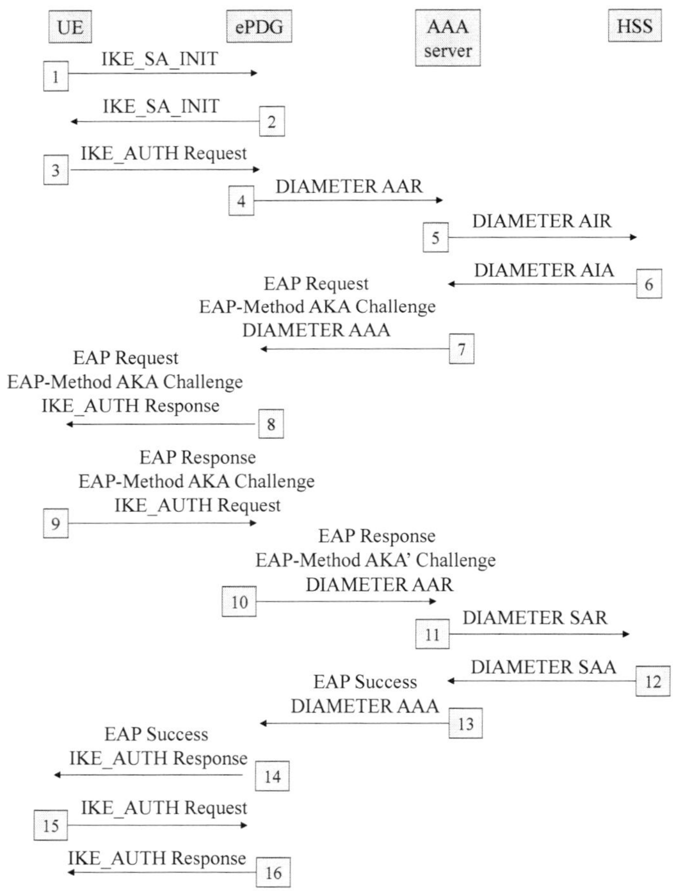

Figure 7.12. *SWu tunnel establishment procedure*

1) and 2) The two IKE_SA_INIT messages are used to negotiate the IKEv2 security association algorithms and to exchange D-H public values and random numbers (Nonce).

3) The mobile transmits the first message Request of the IKE_AUTH phase containing SWu tunnel configuration proposals in the SA block, its identity in the IDi block and access point name (APN) information in the IDr block.

The mobile does not transmit the AUTH block in order to warn the ePDG entity that it wishes to use the IKEv2 message to transport the EAP-AKA method.

The identity of the mobile conforms to the network access identifier (NAI) format containing the international mobile subscriber identity (IMSI) during the first authentication, or during the following authentications, a pseudonym or an identifier for the rapid renewal of authentication.

The mobile transmits the CP block (CFG_REQUEST) in the IKE_AUTH Request message to obtain its IPv4 and/or IPv6 address, and possibly the IP address of the PGW entity, in the case where the mobility is managed by the mobile.

4) The ePDG entity transmits to the AAA server the AAR (Authenticate and Authorize Request) DIAMETER message containing the identity of the mobile and the information relating to the APN.

NAI analysis allows the AAA server to distinguish between authentication for trusted Wi-Fi access based on the EAP-AKA’ mechanism and authentication for untrusted Wi-Fi access based on the AKA mechanism.

5) The AAA server requests the home subscriber server (HSS) for mobile cryptographic data in the AIR (Authentication-Information-Request) DIAMETER message.

The HSS entity generates the RAND sequence and creates the seals (RES and AUTN) and the keys (CK and IK) from the Ki key and the RAND sequence.

6) The HSS entity passes the authentication vectors to the AAA server in the AIA (Authentication-Information-Answer) DIAMETER message.

The AAA server derives the CK and IK to generate the master session key (MSK).

7) The AAA server initiates the authentication procedure with the message EAP Request/EAP-Method AKA Challenge containing the AUTN

and RAND parameters. This message is transmitted in the AAA (Authenticate and Authorize Answer) DIAMETER message.

8) The ePDG entity transfers the message EAP Request/EAP-Method AKA Challenge in the message IKE_AUTH Response containing its identity, certificate and signature.

The mobile verifies the signature of the message IKE_AUTH Response with the public key of the ePDG entity retrieved from its certificate.

The mobile generates the RES, AUTN, CK and IK parameters from the Ki key and the received RAND sequence and compares the received AUTN seal with the locally calculated one. If both seals are identical, the AAA server is authenticated.

The mobile derives both CK and IK to generate the master key (MSK).

9) The mobile transmits the message EAP Response/EAP-Method AKA Challenge containing the RES seal in the message IKE_AUTH Request.

10) The ePDG entity transfers the message EAP Response/EAP-Method AKA Challenge in the AAR DIAMETER message to the AAA server, which compares the received RES seals respectively from the mobile and the HSS entity. If the two seals are identical, then the mobile is authenticated.

11) The AAA server transmits the SAR (Server-Assignment-Request) DIAMETER message to the HSS entity to register itself.

12) The HSS entity responds to the AAA server with the SAA (Server-Assignment-Answer) DIAMETER message containing the mobile profile. The AAA server verifies that Wi-Fi access is allowed.

13) The AAA server transmits to the ePDG entity the AAA DIAMETER message containing the EAP Success message, the MSK and the mobile profile.

14) The ePDG entity stores the MSK and forwards the EAP Success message into the message IKE_AUTH Response.

15) The mobile generates the message IKE_AUTH Request containing in the AUTH block a seal calculated from its MSK, which allows the authentication of the first message IKE_SA_INIT.

16) The ePDG entity checks the seal and starts the S2b tunnel setup procedure described in Chapter 8.

The ePDG entity responds with the message IKE_AUTH Response containing in the AUTH block a seal calculated from its MSK, which enables authentication of the second message IKE_SA_INIT.

The message IKE_AUTH Response is also used to transfer to the mobile its configuration in the CP block (CFG_REPLY) and the final configuration of the SWu tunnel in the SA block.

The mobile configuration was received from the PGW when establishing the S2b tunnel described in Chapter 8.

7.2.2. Procedure for rapid renewal of authentication

The procedure for rapid renewal of authentication is described in Figure 7.13.

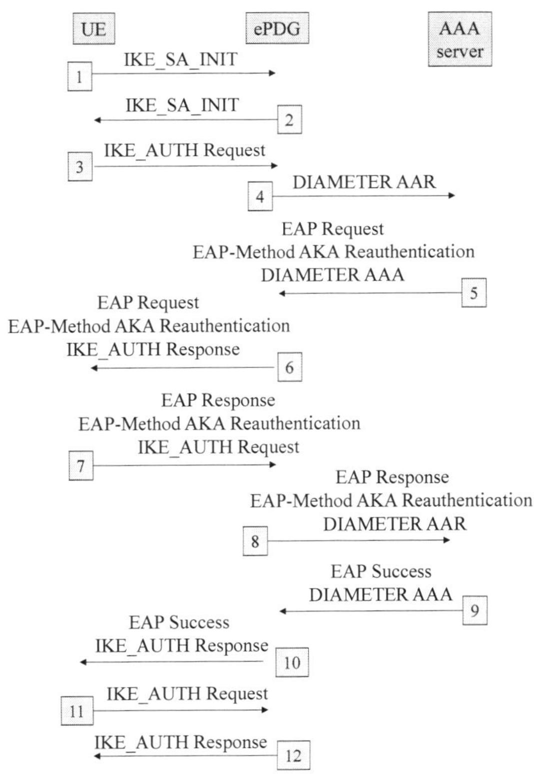

Figure 7.13. *Procedure for rapid renewal of authentication*

Steps 1 to 4 are identical to those described for the establishment of the SWu tunnel in Figure 7.12.

The identifier for rapid renewal of authentication is transmitted in the IDi block of the first message IKE_AUTH Request.

5) The AAA server initiates the procedure for rapid renewal of authentication with the message EAP Request/EAP-Method AKA Reauthentication.

6) The ePDG entity transmits the message IKE_AUTH Response containing its identity, certificate and signature of the IKE_SA_INIT message in the AUTH block.

The message AKA Reauthentication EAP Request/EAP-Method is included to start the EAP procedure on IKEv2.

7) The mobile verifies the signature and responds with the message IKE_AUTH Request and the EAP Response/EAP-Method Reauthentication message containing the mobile seal.

8) The ePDG entity transfers the message EAP Response/EAP-Method Reauthentication to the AAA server.

Steps 9 to 12 are identical to steps 13 to 16 described for the establishment of the SWu tunnel in Figure 7.12.

The new MSK is generated by the AAA server and passed to the ePDG entity and the mobile. This new key is used to authenticate the first two IKE_SA_INIT messages.

8

S2a/S2b Tunnel Establishment

8.1. PMIPv6 mechanism

The PMIPv6 (Proxy Mobile Internet Protocol version 6) mechanism allows a mobile host to keep its original IPv6 address, to maintain its current session or to be reachable when moving, mobility being provided by the network.

The mobile node (MN) is a host that changes network while retaining the home address (HoA) provided by its home network (Figure 8.1).

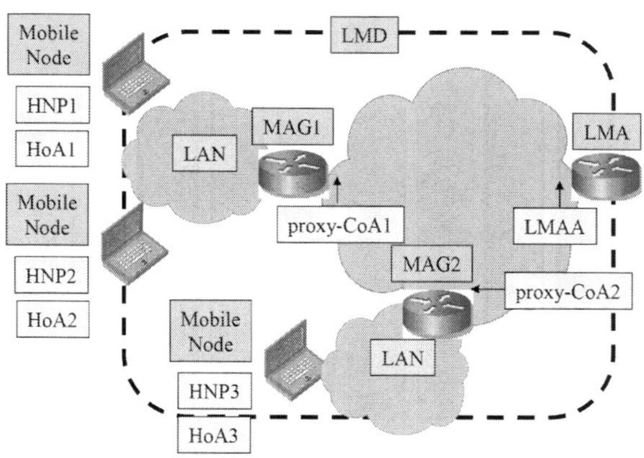

Figure 8.1. *PMIPv6 architecture*

The mobile access gateway (MAG) is integrated into gateway router of the mobile node and provides mobility management for the mobile node connected to its local network (Figure 8.1).

The local mobility anchor (LMA) is built into the router that acts as the home agent (HA) of the mobile node and represents the anchor point for the mobile node (Figure 8.1).

In the case of auto-configuration, the LMA function provides the mobile node with an IPv6 home network prefix (HNP), from which the mobile node builds its HoA.

If not, the MAG function hosts a DHCPv6 server that assigns the HoA to the mobile, built from the IPv6 HNP.

The LMA function registers in the BCE (Binding Cache Entry) table the identity MN-ID of the mobile and the proxy care-of address (CoA) of the MAG of the mobile node.

The tunnel built between the MAG and LMA functions is characterized by the proxy-CoA on the MAG side and the LMA address (LMAA) on the LMA side.

The local mobility domain (LMD) is a set consisting of an LMA function and several MAG functions attached to the LMA function (Figure 8.1).

8.1.1. *Mobility extension*

The mobility extension of the IPv6 header, described in section 9.1.1.1, has defined two types for the MIPv6 mechanism, namely Binding Update and Binding Acknowledgment. These two types are modified for their use adapted to the PMIPv6 mechanism.

The PBU (Proxy Binding Update) extension is the request transmitted by the MAG function to the LMA function to populate the BCE table with the MN-ID identity and the proxy-CoA.

The PBA (Proxy Binding Acknowledgment) extension is the response of the LMA function containing the HNP assigned to the mobile.

The mobility extension can also include the following options:

– Mobile Node Identifier: this option contains the identity of the mobile node that the MAG function retrieved during the authentication of the mobile node;

– Home Network Prefix: this option contains the IPv6 prefix assigned to the mobile node by the LMA function;

– Handoff Indicator: this option indicates that there has been a change of MAG. This option is also used to refresh the BCE table;

– Access Technology Type: this option provides the type of access network to which the mobile node is connected;

– Timestamp: this option provides a timestamp calculated in the number of seconds elapsed since 1 January 1970;

– Mobile Node Link-layer Identifier: this option contains the MAC (Medium Access Control) address of the mobile node;

– Link-local Address: this option contains the proxy-CoA of the MAG function for the PBU extension or the LMAA of the LMA function for the PBA extension.

8.1.2. *Procedures*

8.1.2.1. *Mobile node attachment to the LMA function*

At the end of the connection phase on the local network, the mobile node transmits the ICMPv6 message Router Solicitation in order to retrieve its IPv6 address configuration (Figure 8.2).

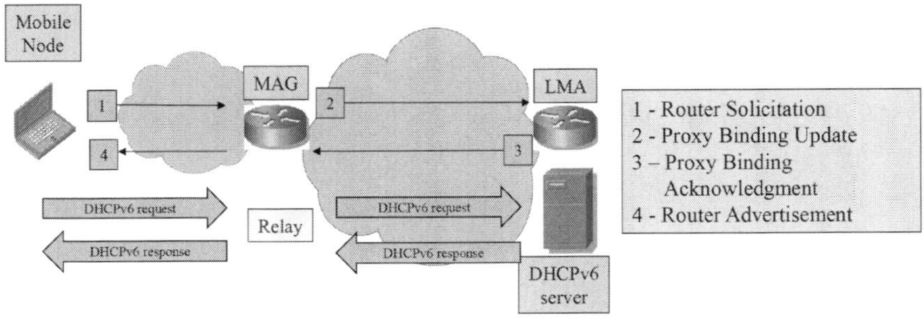

Figure 8.2. *Mobile node attachment to the LMA function IPv6 configuration*

Upon receipt of this message, the MAG function transmits the PBU extension to the LMA function. This extension contains the identity (MN-ID) of the mobile node and the proxy-CoA for tunnel mounting.

The LMA function records in its BCE base the identity (MN-ID) of the mobile node and the proxy-CoA IP of the MAG function.

The LMA function responds to the MAG function with the PBA extension containing the LMAA for tunnel mounting and the HNP for configuring the IPv6 address of the mobile.

The PBU and PBA extensions exchanged between the MAG and LMA functions must be protected by the Internet protocol security (IPSec) by providing the integrity control service.

In the case of an auto-configuration of the IPv6 address by the mobile, the MAG function transfers the prefix HNP to the mobile node in the ICMPv6 message Router Advertisement. The mobile node builds its IPv6 address HoA from the received HNP.

In the opposite case, the MAG function indicates in the ICMPv6 message Router Advertisement that the mobile must acquire its HoA from the DHCPv6 server.

The MAG function must integrate a relay of the DHCPv6 request issued by the mobile node to the DHCPv6 server and the response of the DHCHv6 server. The relayed DHCPv6 request must contain the HNP that the MAG function has received from the LMA function. The DHCPv6 server must use this prefix to assign, in its response, the HoA to the mobile node.

8.1.2.2. *MAG function change*

The moving of the mobile node can result in a change of MAG function, the mobile having to disconnect from the previous MAG (p-MAG) and connect to a new MAG (n-MAG) (Figure 8.3).

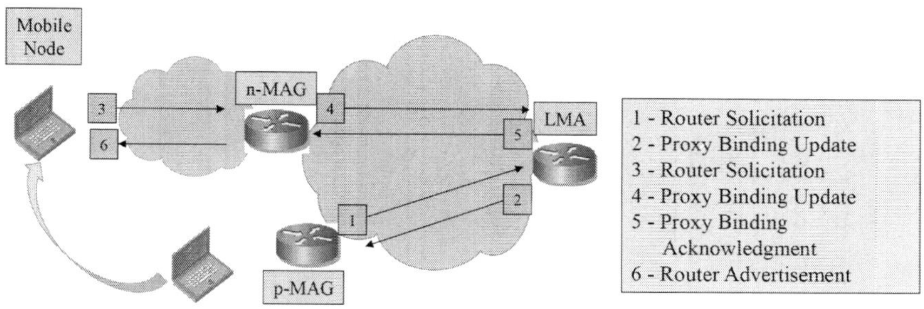

Figure 8.3. *MAG function change*

The p-MAG function warns the LMA function of the mobile disconnection in the PBU extension.

The LMA function removes the proxy-CoA from the p-MAG function from its BCE table, arms a timer to allow the update of the proxy-CoA of the n-MAG function and transmits an acknowledgment to the p-MAG function in the PBA extension.

At the end of the connection of the mobile node, the procedure identical to the previous one is carried out:

– the mobile node transmits the ICMPv6 message Router Solicitation;

– the n-MAG function transmits the PBU extension containing the identity (MN-ID) of the mobile node and the proxy-CoA of the n-MAG function;

– the LMA function transmits the PBA extension containing the LMAA and the HNP;

– the n-MAG function transmits the ICMPv6 message Router Advertisement containing the HNP, which allows the mobile to ensure that its IPv6 address is maintained.

8.1.3. *Application to the 4G mobile network*

8.1.3.1. *Trusted Wi-Fi access*

The LMA and MAG functions are hosted, respectively, by the PGW entity and trusted Wi-Fi access.

The GRE (Generic Routing Encapsulation) protocol constructs the S2a tunnel from a key provided by the trusted Wi-Fi access for the downstream traffic and a key provided by the PGW for the traffic in the upstream direction.

The procedure for establishing the S2a tunnel is described in Figure 8.4 and corresponds to the auto-configuration of the IPv6 address by the mobile.

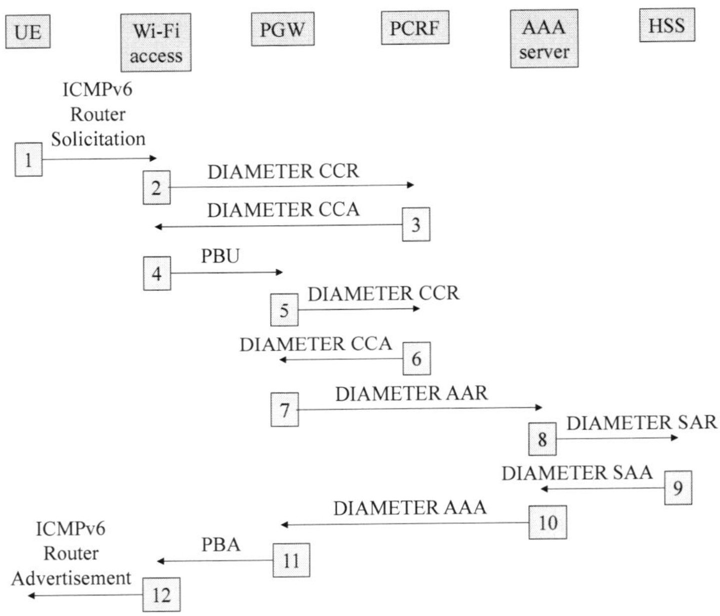

Figure 8.4. *S2a tunnel establishment using PMIPv6 mechanism*

The procedure for S2a tunnel establishment starts when the mobile authentication, described in Chapter 6, is successful.

1) The mobile hands over the ICMPv6 message Router Solicitation to retrieve its IPv6 address configuration.

This message may contain the access point name (APN) that allows Wi-Fi access to determine the IP address of the PGW.

Otherwise, Wi-Fi access uses the default access point name that is passed by the AAA server during mobile authentication.

2) The Wi-Fi access transmits to the PCRF entity the DIAMETER message CCR (Credit-Control-Request) containing the mobile profile received from the AAA server during the authentication, to obtain the authorization for the opening of the default bearer.

The PCRF compares with the rules defined for the network and stored in the SPR (Subscription Profile Repository) database.

3) The PCRF responds to Wi-Fi access with the DIAMETER message CCA (Credit-Control-Answer) containing the rules to apply to the default bearer.

4) Wi-Fi access transmits to the PGW entity the PBU extension containing the following parameters: MN-NAI, Lifetime, Access Technology Type, APN, GRE key for downlink traffic, Charging Characteristics and Additional Parameters.

5) The PGW entity sends the PCRF entity the CCR DIAMETER message to obtain the default bearer characteristics.

6) The PCRF entity responds to the PGW entity with the CCA DIAMETER message containing the rules to apply to the default bearer (filter parameter, charging mode).

7) The PGW entity sends the AAA server the DIAMETER message AAR (Authenticate and Authorize Request) to communicate its identity and the access point name for the connection.

8) The AAA server sends the HSS entity the DIAMETER message SAR (Server-Assignment Request) to transfer the information received from the PGW entity.

9) The HSS entity responds to the AAA server with the DIAMETER message SAA (Server-Assignment-Answer) that contains the mobile profile:

– the access point name (APN);

– QoS (Quality of Service) characteristics for each default bearer to be established.

10) The AAA server responds to the PGW entity with the DIAMETER message AAA (Authenticate and Authorize Answer) containing the information received from the HSS entity.

The PGW will use the mobile profile received from the AAA server if these parameters were not provided by the PCRF.

11) The PGW entity responds to the Wi-Fi access point with the PBA extension containing the following parameters: MN NAI, Lifetime, UE Address Info, GRE key for uplink traffic, Charging ID and Additional Parameters.

12) Wi-Fi access responds to the mobile with the ICMPv6 message Router Advertisement containing the mobile configuration parameters (IPv6 prefix, IP address of the DNS server).

8.1.3.2. *Untrusted Wi-Fi access*

The LMA and MAG functions are hosted by the PGW and ePDG entities, respectively.

The GRE protocol constructs the S2b tunnel from a key provided by the ePDG entity for downstream traffic and a key provided by the PGW for upstream traffic.

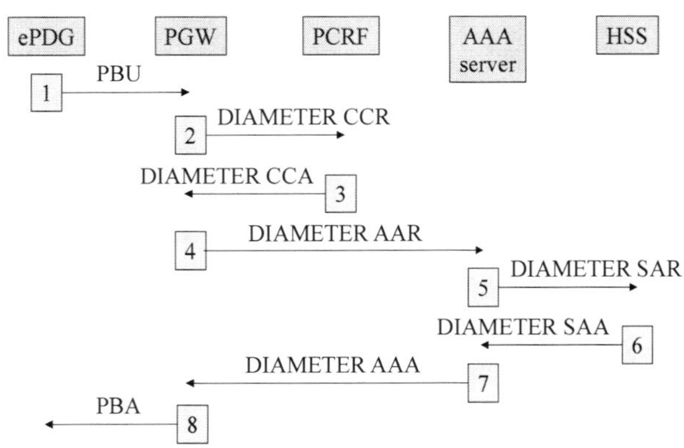

Figure 8.5. *S2b tunnel establishment using PMIPv6 mechanism*

The procedure for establishing the S2b tunnel starts during the SWu tunnel establishment procedure described in Chapter 7 (Figure 8.5).

1) The ePDG entity transmits to the PGW entity the PBU extension containing the following fields: MN-NAI, Lifetime, APN, Access Technology Type, GRE key for downlink traffic, UE Address Info, Charging Characteristics and Additional Parameters.

2) The PGW entity sends the PCRF entity the CCR DIAMETER message to obtain the mobile traffic profile.

3) The PCRF entity responds to the PGW entity with the CCA DIAMETER message containing the rules to be applied to the default bearer (APN-AMBR rate parameters and QoS).

4) The PGW entity sends the AAA server the AAR DIAMETER message to communicate its identity and the access point name for the connection.

5) The AAA server transmits to the HSS entity the SAR DIAMETER message to transfer the information received from the PGW entity.

6) The HSS entity responds to the AAA server with the SAA DIAMETER message to transfer the information received from the PGW entity.

7) The AAA server responds to the PGW entity with the AAA DIAMETER message containing the information received from the HSS entity. The mobile profile is taken into account if the PCRF did not provide the information in step 3.

8) The PGW entity responds to the ePDG entity with the PBA extension containing the following fields: MN-NAI, UE Address Info, GRE Key for uplink traffic and Charging ID.

The ePDG entity completes the SWu tunnel establishment procedure described in Chapter 6.

8.2. GTPv2 mechanism

The GTPv2 (GPRS Tunneling Protocol version 2) mechanism comprises the GTPv2-C (Control) signaling that manages the S2a or S2b tunnel and the GTP-U (User) protocol for building the S2a or S2b tunnel.

The GTPv2-C protocol allows the establishment or closure of the mobile context and the bearers of the mobile streams (Table 8.1).

Message type	Request	Response
Context management	CREATE/DELETE SESSION REQUEST	CREATE/DELETE SESSION RESPONSE
Bearer management	CREATE/MODIFY/DELETE BEARER REQUEST	CREATE/MODIFY/DELETE BEARER RESPONSE

Table 8.1. *GTPv2-C messages*

The messages CREATE SESSION REQUEST/RESPONSE allow the creation of the context and possibly the default bearer.

The messages CREATE BEARER REQUEST/RESPONSE allow the creation of default and dedicated bearers.

The context is a collection of mobile-related information including identifiers, location, security and bearer characteristics.

The tunnel is identified by the Tunnel Endpoint Identifier (TEID) carried by the GTP-U protocol, tunnel end IP addresses and UDP port numbers. The entity receiving the traffic data determines the value of the TEID parameter that the sending entity is to use.

8.2.1. *Trusted Wi-Fi access*

The GTP-U protocol constructs the S2a tunnel from a TEID provided by the trusted Wi-Fi access for the downstream traffic and a TEID provided by the PGW entity for the upstream traffic.

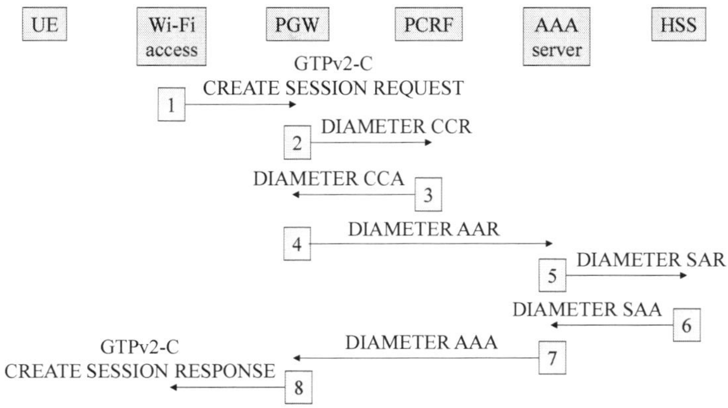

Figure 8.6. *S2a tunnel establishment using GTPv2 mechanism*

The S2a Tunneling procedure, shown in Figure 8.6, starts during the mutual authentication procedure for the mobile and AAA server detailed in Chapter 6.

1) The Wi-Fi access transmits to the PGW entity the GTPv2C message CREATE SESSION REQUEST containing the following fields: IMSI, APN, RAT type, PDN Type, PDN Address, Bearer Identity EPS, Default EPS QoS Bearer, AP Address, AP TEID, APN-AMBR, Charging Characteristics and Additional Parameters.

2) The PGW entity sends the PCRF entity the CCR DIAMETER message to obtain the default bearer characteristics. The PCRF entity can change the value of the APN-AMBR.

3) The PCRF entity responds to the PGW entity with the CCA DIAMETER message containing the rules to be applied to the default bearer (QoS parameters, filter parameter, charging mode).

4) The PGW entity sends the AAA server the AAR DIAMETER message to communicate its identity and the access point name for the connection.

5) The AAA server transmits to the HSS entity the SAR DIAMETER message to transfer the information received from the PGW entity.

6) The HSS entity responds to the AAA server with the SAA DIAMETER message.

7) The AAA server responds to the PGW entity with the AAA DIAMETER message.

8) The PGW entity responds to trusted Wi-Fi access with the GTPv2C message CREATE SESSION RESPONSE, containing the following fields: PGW Address, PGW TEID, PDN Type, PDN Address, EPS Bearer Identity, EPS Bearer QoS, APN-AMBR and Additional Parameters.

The trusted Wi-Fi access completes the authentication procedure (EAP Success message) by providing the elements of its configuration contained in the Additional Parameters field.

8.2.2. *Untrusted Wi-Fi access*

The GTP-U protocol constructs the S2b tunnel from a TEID provided by the ePDG entity for downstream traffic and a TEID provided by the PGW entity for upstream traffic.

The procedure for setting the S2a bearer resumes that described for the PMIPv6 mechanism with the following modifications:

The PBU message in step 1 is replaced by the CREATE SESSION REQUEST message containing the following fields: IMSI, APN, RAT type, TEID ePDG, PDN Type, PDN Address, Bearer Identity EPS, EPS QoS Bearer, ePDG Address, APN-AMBR and Additional Parameters.

The PBA message in step 8 is replaced by the CREATE SESSION RESPONSE message containing the following fields: PDN GW Address, PDN GW TEID, PDN Type, PDN Address, Bearer Identity EPS, EPS Bearer QoS, APN-AMBR and Charging ID.

8.3. MIPv4 FA mechanism

8.3.1. *Components of mobility*

The mobile node is a host that changes network while retaining the HoA of its home network. When attached to a foreign network, it acquires an additional CoA (Figure 8.7).

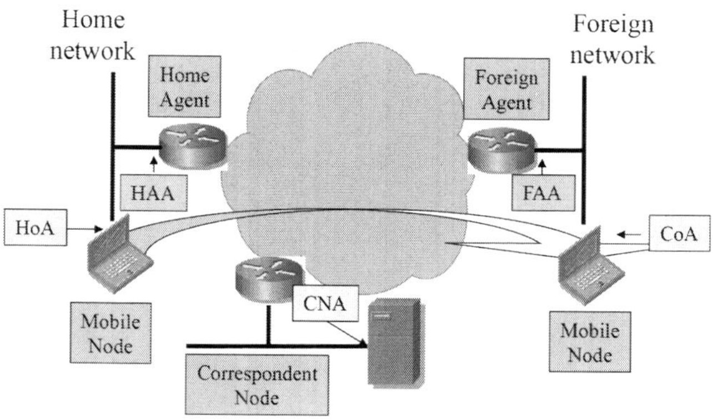

Figure 8.7. *Components of mobility*

The home agent (HA) is the entity of the originating network to which the mobile node must register when it attaches to a foreign network. The role of the home agent is to intercept the received packets and send them back in a tunnel to the mobile node. The HAA is that of the home agent interface on the home network of the mobile node (Figure 8.7).

The foreign agent (FA) is the entity of the network visited by the mobile node. It ends the tunnel and delivers the packets to the mobile node. The foreign agent address (FAA) is the gateway address of the mobile node in the visited network (Figure 8.7).

The correspondent node (CN) is the host that exchanges packets with the mobile node. Its address is noted CNA (Correspondent Node Address) (Figure 8.7).

8.3.2. *Foreign agent discovery*

The foreign agent discovery uses ICMPv4 (Internet Control Message Protocol) messages, Router Solicitation and Router Advertisement.

The Agent Advertisement message is formed by including an extension, Mobility Agent Advertisement in an ICMPv4 message Router Advertisement. The Agent Advertisement message informs the mobile nodes of the capabilities of the foreign agent and provides the CoA.

8.3.3. *Registration*

Registration allows the mobile node to communicate its CoA to the home agent. If the CoA is obtained from the Agent Advertisement message, registration takes place via the foreign agent. If the CoA is obtained from a DHCP server, and if the R bit of the Agent Advertisement message is set to ZERO, then the mobile node can register directly with the home agent.

The registration messages are encapsulated by a UDP header. The number of the destination port (respectively the source port) is equal to 434 for the message Registration Request transmitted by the mobile node (respectively for the response Registration Reply of the home agent).

8.3.4. *Procedure*

Foreign agents announce their presence using ICMPv4 message Agent Advertisement. A mobile node may optionally request an ICMPv4 message Agent Advertisement through a foreign agent solicitation message.

A mobile node receives these Agent Advertisement messages and determines whether it is on its home network or on a foreign network.

When the mobile node detects that it is located on its home network, it operates without the mobility services.

When a mobile node detects that it has moved to a foreign network, it obtains a CoA, determined from the Agent Advertisement message, or through a DHCP server.

The mobile node then registers its CoA with its home agent through the exchange of Registration Request and Registration Reply messages, possibly via a foreign agent.

When the home agent receives and accepts the Registration Request message, it broadcasts a gratuitous ARP (Address Resolution Protocol) message to update the ARP table of the hosts on its network, matching the Ethernet address of the home agent with the HoA.

Similarly, the home agent provides the ARP proxy function to respond to the ARP requests that it receives from hosts in the network regarding the HoA.

Packets sent by the correspondent node to the HoA are intercepted by the home agent. The home agent establishes a tunnel and transfers the packets to the CoA. If the CoA is obtained by the Agent Advertisement message, the end of the tunnel is constituted by the foreign agent that delivers the packets to the mobile node (Table 8.2 and Figure 8.8). In the case where the CoA is obtained by a DHCP server, the mobile node constitutes the tunnel termination.

The correspondent node sends packets to the mobile node	
Packet source address	CNA
Packet destination address	HoA
The home agent establishes the tunnel with the mobile node	
Tunnel source address	HAA
Tunnel destination address	CoA
Packet source address	CNA
Packet destination address	HoA
The foreign agent transfers the packets to the mobile node	
Packet source address	CNA
Packet destination address	HoA

Table 8.2. *Data transfer: CN to MN*

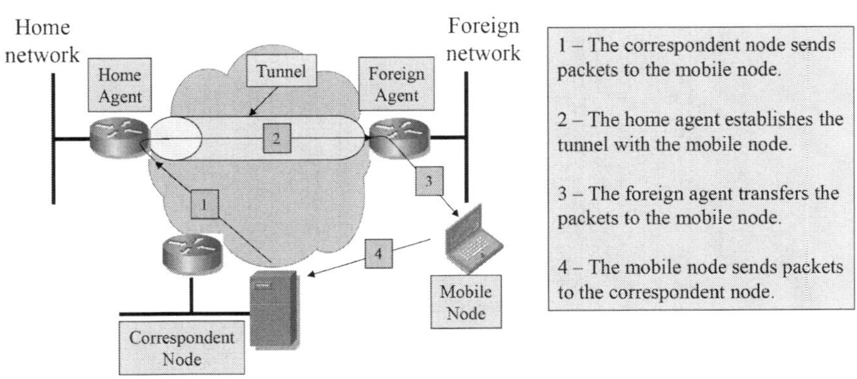

Figure 8.8. *Data transfer*

In the opposite direction, packets sent by the mobile node are usually delivered directly to the corresponding node, without necessarily passing through the home agent (Figure 8.8).

If the mobile node returns to its home network, it unsubscribes with its home agent, by exchanging Registration Request and Registration Reply messages.

When the home agent receives and accepts the Registration Request message, it stops providing the ARP proxy function.

Upon receipt of the Registration Reply message, the mobile node transmits a gratuitous ARP, showing the correspondence between the data-link layer address of the mobile node and its HoA.

8.3.5. *Application to the 4G mobile network*

The home agent (HA) and foreign agent (FA) functions are hosted, respectively, by the PGW entity and trusted Wi-Fi access.

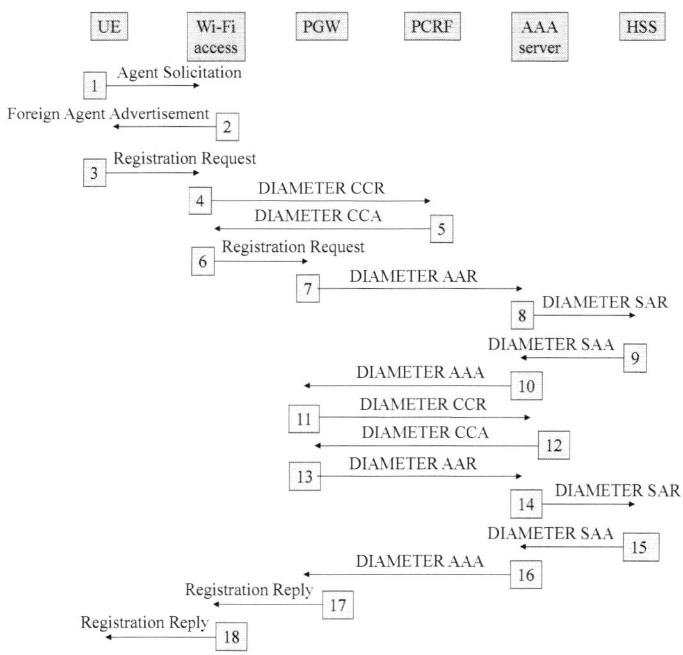

Figure 8.9. *S2a tunnel establishment using MIPv4 FA mechanism*

The procedure for setting up the S2a tunnel, described in Figure 8.9, starts after the mutual authentication procedure for the mobile and the AAA server detailed in Chapter 6.

1) The mobile transmits the ICMPv4 message Agent Solicitation.

2) Wi-Fi access responds to the mobile with the ICMPv4 message Foreign Agent Advertisement, containing the CoA of the foreign agent.

3) The mobile transmits the Registration Request message containing the following fields: MN-NAI, Lifetime and APN.

4) Wi-Fi access sends the PCRF entity the CCR DIAMETER message containing the mobile profile received from the AAA server during authentication, to obtain authorization to open the default bearer.

The PCRF may modify the received parameters if the rules defined for the network and stored in the SPR database are different.

5) The PCRF responds to Wi-Fi access with the CCA DIAMETER message containing the rules to apply to the default bearer.

6) Trusted Wi-Fi access transfers the Registration Request message to the PGW entity.

7) The PGW entity sends the AAR DIAMETER message to the AAA server to retrieve the mobile profile.

8) The AAA server transmits the SAR DIAMETER message to the HSS entity to retrieve the profile of the mobile.

9) The HSS entity responds to the AAA server with the SAA DIAMETER message containing the mobile profile.

10) The AAA server transmits to the PGW entity the AAA DIAMETER message containing the profile of the mobile.

11) The PGW entity sends the PCRF entity the CCR DIAMETER message to obtain the default bearer characteristics. The PCRF can change the value of the aggregate maximum bearer rate (APN-AMBR).

12) The PCRF entity responds to the PGW entity with the CCA DIAMETER message containing the rules to be applied to the default bearer (QoS parameters, filtering parameters, charging mode).

13) The PGW entity sends the AAA server the AAR DIAMETER message to communicate its identity and the access point name for the connection.

14) The AAA server transmits to the HSS entity the SAR DIAMETER message to transfer the information received from the PGW.

15) The HSS entity responds to the AAA server with the SAA DIAMETER message.

16) The AAA server responds to the PGW entity with the AAA DIAMETER message.

17) The PGW entity responds to the trusted Wi-Fi access with the Registration Reply message containing the following fields: MN-NAI, Home Address (HoA), Home Agent Address (HAA) and Lifetime.

18) The trusted Wi-Fi access transfers the Registration Reply message to the mobile that retrieves its HoA.

9

S2c Tunnel Establishment

9.1. MIPv6 mechanism

The MIPv6 (Mobile Internet Protocol version 6) mechanism allows a moving host to keep its original IPv6 address, in order to maintain its current session or to be reachable on the move, mobility being managed by the host.

The mobile node (MN) is a host that changes network while retaining the home address (HoA) of its home network. When attached to a foreign network, it acquires an additional care-of address (CoA) (Figure 9.1).

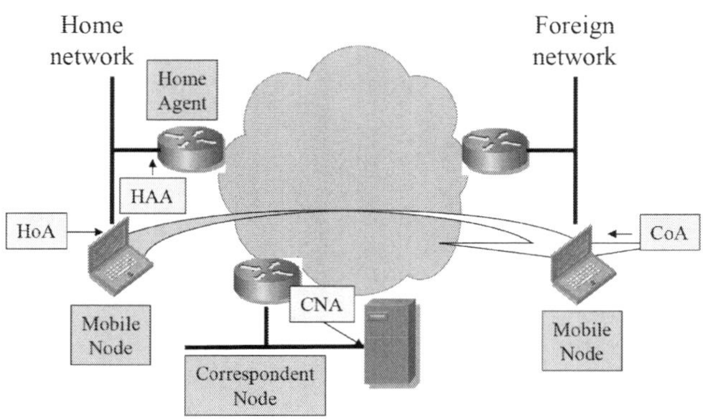

Figure 9.1. *Components for MIPv6 mechanism*

The home agent (HA) is the entity of the originating network from which the mobile node must register when it attaches to a foreign network. The role of the host agent is to intercept the received packets and send them back in a tunnel to the mobile node. The home agent address (HAA) is that of the interface of the home agent on the home network of the mobile node (Figure 9.1).

The correspondent node (CN) is the host that exchanges packets with the mobile node. Its address is noted CNA (Correspondent Node Address) (Figure 9.1).

IPv6 mobility implements packet routing optimization between the mobile node and the correspondent node. The systematic routing of the packets exchanged via the home agent is simple to implement. On the other hand, if the mobile node is moving away from its home network and communicating with a correspondent node close to it, then it is more efficient to communicate directly rather than through the home agent.

9.1.1. *IPv6 header extensions*

9.1.1.1. *Mobility extension*

The mobility extension, described in Figure 9.2, is attached to the IPv6 header and allows the exchange of information between the mobile node, on the one hand, and the correspondent node or the home agent, on the other.

```
 0                   1                   2                   3
 0 1 2 3 4 5 6 7 8 9 0 1 2 3 4 5 6 7 8 9 0 1 2 3 4 5 6 7 8 9 0 1
+-+-+-+-+-+-+-+-+-+-+-+-+-+-+-+-+-+-+-+-+-+-+-+-+-+-+-+-+-+-+-+-+
| Payload Proto | Header Length |    MH Type    |   Reserved    |
+-+-+-+-+-+-+-+-+-+-+-+-+-+-+-+-+-+-+-+-+-+-+-+-+-+-+-+-+-+-+-+-+
|           Checksum            |                               |
+-+-+-+-+-+-+-+-+-+-+-+-+-+-+-+-+                               |
|                                                               |
//                         Message Data                        //
|                                                               |
+-+-+-+-+-+-+-+-+-+-+-+-+-+-+-+-+-+-+-+-+-+-+-+-+-+-+-+-+-+-+-+-+
```

Figure 9.2. *Mobility extension format*

This information relates to the registration of the mobile node with the correspondent node or the host agent, for the association of addresses HoA and CoA.

The Next Header field in the header preceding the mobility extension has a value of 135.

Proto Payload: this field, coded on one byte, has a value of 59, indicating termination of extensions chaining.

Header Length: this field, coded on one byte, defines the size of the extension.

MH (Mobility Header) Type: this field, coded on one byte, defines the type of mobility extension:

– 0) Binding Refresh Request. This extension is transmitted by the correspondent node or by the home agent to the mobile node in order to update the link between the HoA and CoA.

– 1) Home Test Init (HoTI). This extension initializes the Return Routability procedure on the indirect path between the mobile node and the correspondent node.

– 2) Care-of Test Init (CoTI). This extension initializes the Return Routability procedure on the direct path between the mobile node and the correspondent node.

– 3) Home Test (HoT). This extension is the answer to the HoTI extension. It contains the cryptographic information (Home Keygen Token).

– 4) Care-of Test (CoT). This extension is the answer to the HoTI extension. It contains the cryptographic information (Care-of Keygen Token).

– 5) Binding Update. This extension is transmitted by the mobile node to link the HoA contained in the Home Address option of the Destination extension to the CoA contained in the source address of the IPv6 header or in the Alternate Care-of Address option.

– 6) Binding Acknowledgment. This extension is passed by the correspondent node or by the home agent to acknowledge receipt of the Binding Update message.

– 7) Binding Error. This message is transmitted by the correspondent node or by the home agent if the Binding Update message is incorrect.

Checksum: this field, coded on two bytes, contains the checksum calculated on the pseudo-header and the mobility extension.

Message Data: this variable size field contains the data corresponding to the type of mobility extension.

The mobility extension can also include the following options:

– Pad1: this option is used to insert a padding byte;

– PadN: this option is used to insert several bytes of padding;

– Binding Refresh Advice: this option is associated with the mobility extension Binding Acknowledgment passed by the home agent. It specifies the value of the timer used by the mobile node to update its registration;

– Alternate Care-of Address: this option is associated with the mobility extension Binding Update. It specifies the CoA if it cannot be deduced from the source address of the IPv6 header;

– Nonce Indices: this option is associated with the mobility extension Binding Update passed to the correspondent node. It contains random numbers (nonce) needed for calculating cryptographic information (Home Keygen Token, Care-of Keygen Token);

– Binding Authorization Data: this option is associated with the mobility extensions Binding Update and Binding Acknowledgment. It contains cryptographic information from which the destination can verify that the message originates from a node with which the Return Routability procedure has occurred.

9.1.1.2. *Destination extension*

The Home Address option of the Destination extension indicates the HoA of the mobile node.

The Destination extension is used for the direct transfer of data from the mobile node to the correspondent node.

The mobile node cannot use the HoA as the source address because the router of the foreign network can delete the packet if the source does not belong to the local network.

The mobile node is therefore obliged to retain the CoA as the source address.

In reception, the correspondent node must replace the CoA with the address HoA to reconstitute the socket.

9.1.1.3. *Routing extension*

The Routing extension (type 2) contains the HoA of the mobile node.

The Routing extension is used for the direct transfer of data from the correspondent node to the mobile node.

The correspondent node transmits the packet to the CoA destination address of the mobile node.

Upon receiving the packet, the mobile node replaces the CoA with the HoA of the Routing (type 2) extension to reconstruct the socket.

9.1.2. *ICMPv6 messages*

9.1.2.1. *Message Home Agent Address Discovery Request*

The message Home Agent Address Discovery Request is transmitted by the mobile node to the home agent. The source address is the CoA of the mobile node. The destination address is the anycast address constructed from the HoA of the mobile node. The value of the Type field of the ICMPv6 message is 144.

9.1.2.2. *Message Home Agent Address Discovery Reply*

The message Home Agent Address Discovery Reply is transmitted by the home agent in response to the previous message. It contains a list of HAA. The value of the Type field of the ICMPv6 message is 145.

9.1.2.3. *Message Mobile Prefix Solicitation*

The message Mobile Prefix Solicitation is transmitted by the mobile node to the host agent in order to update the configuration of its HoA. The source address is the CoA of the mobile node. The destination address is the HAA. The Destination extension must be inserted. The value of the Type field of the ICMPv6 message is 146.

9.1.2.4. *Message Mobile Prefix Advertisement*

The message Mobile Prefix Advertisement is transmitted by the home agent either in response to the previous message or in an unsolicited manner. In both cases, the destination address is the CoA of the mobile node. The Routing extension (Type 2) must be inserted. The value of the ICMPv6 message type field is 147. The ICMPv6 message incorporates the Prefix Information option.

9.1.2.5. *ND protocol modifications*

The changes to the ICMPv6 ND (Neighbor Discovery) protocol are as follows:

– the RA (Router Advertisement) message is modified;

– the Prefix Information option is modified;

– the Advertisement Interval option is created;

– the Home Agent Information option is created.

The H (Home Agent) flag is introduced in the Router Advertisement message to allow a home agent to discover other home agents on the home network.

The RA message uses the LINK-LOCAL address as the source address. The Prefix Information option includes the prefix used to set up the GLOBAL UNICAST address.

The home agent listening for Router Advertisement messages from other home agents cannot get their GLOBAL UNICAST address. The Prefix Information option introduces an R (Router Address) flag to signify that the announced prefix is actually a GLOBAL UNICAST address.

The Advertisement Interval option is passed in the Router Advertisement message to set the frequency of sending messages.

The Home Agent Information option is associated with the Router Advertisement message sent by the home agent. It specifies the level of preference of the home agent and its lifetime. The preference level is used by the home agent to order the home agent list transmitted in the message Home Agent Address Discovery Reply.

9.1.3. *Procedures*

9.1.3.1. *Attachment of the mobile node to the home agent*

When the mobile node detects a network change, it performs the DAD (Duplicate Address Detection) procedure with its LINK-LOCAL address.

It then discovers the network prefix by ICMPv6 messages, Router Solicitation and Router Advertisement; builds its CoA and verifies its uniqueness (Figure 9.3).

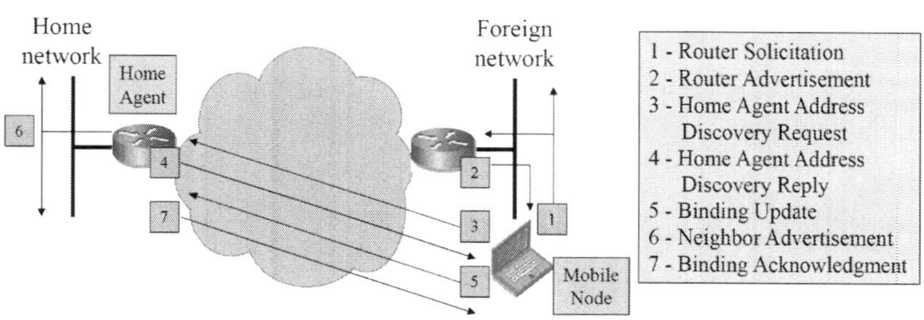

Figure 9.3. *Attachment of the mobile node to the home agent*

Sometimes, when the mobile node has to send a Binding Update message to its home agent to register its CoA, it may not know the HAA.

In this case, the mobile node may attempt to discover the address of a home agent by sending the message Home Agent Address Discovery Request using the anycast address corresponding to the prefix of its HoA.

The host agent that receives this request message returns the message Home Agent Address Discovery Reply containing the list of HAA (Figure 9.3).

The mobile node, after receiving the message Home Agent Address Discovery Reply, can then send the Binding Update message to one of the HAA.

The mobile node may attempt to register at each of these addresses, in the order they appear in the message Home Agent Address Discovery Reply, until its registration is acknowledged by receiving the Binding Acknowledgment message (Figure 9.3).

The home agent broadcasts the Neighbor Advertisement message to refresh the neighbor table of the hosts of the originating network, for which the HoA of the mobile node is associated with the Ethernet address of the home agent (Figure 9.3).

9.1.3.2. *Data transfer*

The transfer of packets between the correspondent node and the mobile node takes place initially through the home agent (Figure 9.4).

Packets from the correspondent node to the mobile node are intercepted by the home agent.

Packet source address: CNA
Packet destination address: HoA

The home agent encapsulates the packets received by a new IPv6 header to transfer them to the mobile node.

Tunnel source address: HAA
Tunnel destination address: CoA
Packet source address: CNA
Packet destination address: HoA

Packets from the mobile node to the correspondent node are forwarded in a tunnel to the home agent.

Packet source address: HoA
Packet destination address: CNA
Tunnel source address: CoA
Tunnel destination address: HAA

The home agent deletes the tunnel and forwards the packets to the correspondent node.

Packet source address: HoA
Packet destination address: CNA

The mobile node then implements the direct transfer procedure, initially initializing the Return Routability procedure. It then transmits the Binding Update message to the correspondent node to create the link between the CoA and HoA (Figure 9.4).

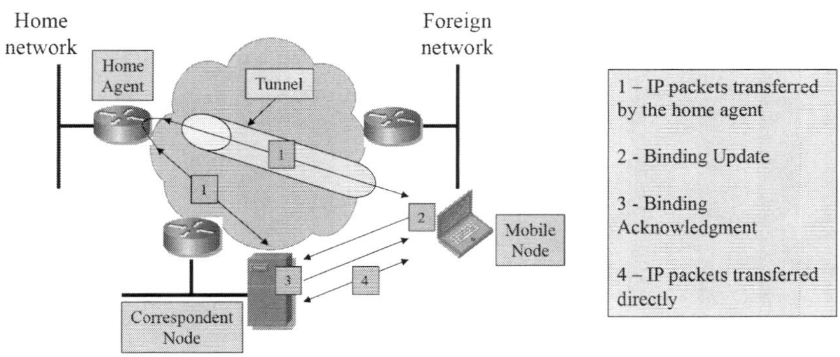

Figure 9.4. *Data transfer*

The transfer of packets between the correspondent node and the mobile node can then be carried out directly (Figure 9.4).

Packets from the correspondent node to the mobile node carry the following addresses:

Packet source address: CNA
Packet destination address: CoA
Routing (type 2) extension: HoA

Packets from the mobile node to the correspondent node carry the following addresses:

Packet source packet: CoA
Packet destination address: CNA
Destination extension: HoA

9.1.3.3. *Local network change*

When the mobile node communicating with the correspondent node changes a foreign network, it repeats the procedure in order to build its new CoA, through the messages Router Solicitation and Router Advertisement (Figure 9.5).

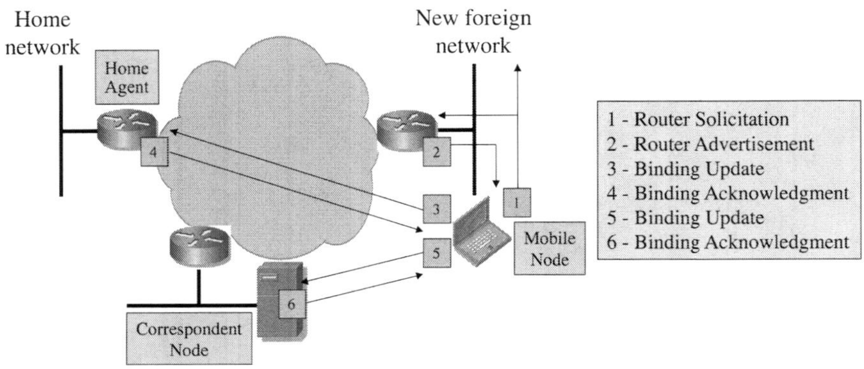

Figure 9.5. *Network change of the mobile node*

The mobile node resumes the operation of registering its CoA with its home agent by the messages Binding Update and Binding Acknowledgment (Figure 9.5)

When the mobile node has registered with its home agent, it triggers a registration at the correspondent node to update the CoA by the messages Binding Update and Binding Acknowledgment. This registration is preceded by the Return Routability procedure (Figure 9.5).

9.1.3.4. *Return of the mobile node to the host network*

A mobile node detects that it has returned to its home network when it detects its prefix in a Router Advertisement message. To be able to send and receive packets using its HoA, the mobile node must send a Binding Update message to its home agent to warn it to stop intercepting packets (Figure 9.6).

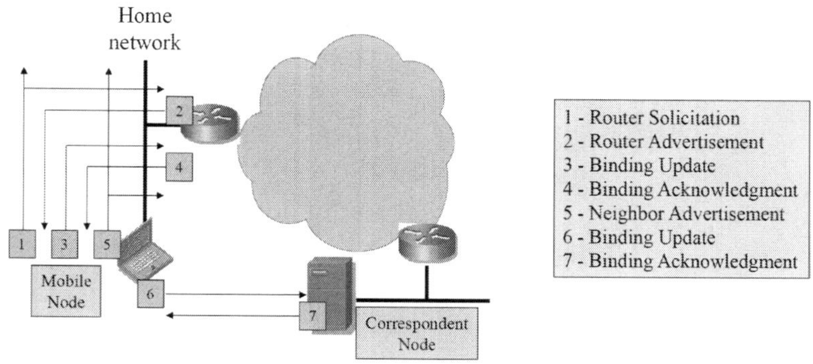

Figure 9.6. *Return of the mobile node to the host network*

By processing the Binding Update message, the home agent will stop responding to Neighbor Solicitation messages regarding the HoA of the mobile node. Upon receiving the Binding Acknowledgment message, the mobile node broadcasts the Neighbor Advertisement message to update the neighbor table of the hosts of the home network (Figure 9.6).

The mobile node renews the operation of registering its CoA with the correspondent node by the exchange of messages Binding Update and Binding Acknowledgment. This registration is preceded by the Return Routability procedure, limited to the exchange of HoTI and HoT messages (Figure 9.6).

9.1.3.5. *Return Routability procedure*

The Return Routability procedure allows the correspondent node to ensure that the mobile node is in fact accessible to its CoA and HoA. This assurance allows the correspondent node to accept the Binding Update message sent by the mobile node for the purpose of establishing a direct transfer.

The HoTI and CoTI messages are transmitted simultaneously by the mobile node to the correspondent node (Figure 9.7):

– HoTI message passes through the home agent;

– CoTI message is sent directly to the correspondent node.

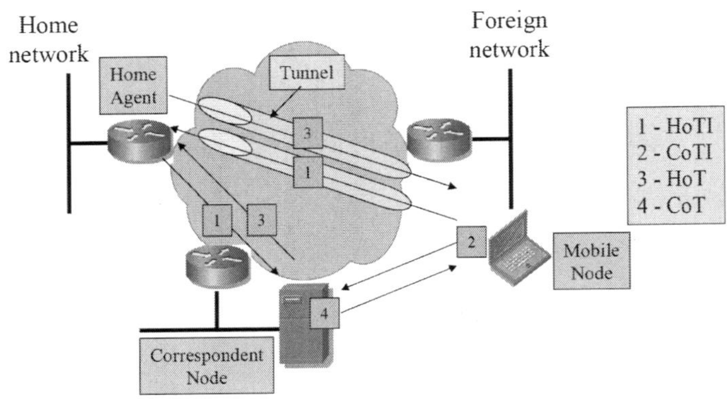

Figure 9.7. *Return Routability procedure*

The HoT and CoT messages are the responses of the correspondent node to the mobile node (Figure 9.7):

– HoT message contains the cryptographic information Home Keygen Token computed from the HoA, from a random number (nonce) and from the secret key Kcn of the correspondent node:

First 64 bits HMAC_SHA1 (Kcn, (HoA | nonce | 0x00))

– CoT message contains the cryptographic information Care-of Keygen Token calculated from the address CoA, a random number (nonce) and the secret key Kcn of the correspondent node:

First 64 bits HMAC_SHA1 (Kcn, (CoA | nonce | 0x01))

Following the procedure, the mobile node has the data it needs to build a Kbm secret key by hashing the received data:

Kbm = SHA1 (Home Keygen Token | Care-of Keygen Token)

The Binding Update message transmitted by the mobile node directly to the correspondent node contains the following cryptographic information:

First 96 bits HMAC_SHA1 (Kbm, (CoA | CNA | Binding Update))

When the correspondent node has validated the received Binding Update message, it responds with the Binding Acknowledgment message with the following cryptographic information:

First 96 bits HMAC_SHA1 (Kbm, (CoA | CNA | Binding Acknowledgement))

The procedure is based on the assumption that no intruder can listen to both HoT and CoT messages, these messages using two different paths to join the mobile node. Interception remains possible if the malicious node is connected to the network of the correspondent node.

The procedure is based on the shared Kbm secret that needs to be refreshed. Refreshment is left to the initiative of the correspondent node. An association change request Binding Update is denied through the Binding Error message. The mobile node must then restart the Return Routability procedure.

9.2. DSMIPv6 mechanism

The MIPv6 mechanism was designed for a mobile connection to an IPv6 network. The DSMIPv6 (Dual-Stack Mobile IP version 6) mechanism also takes into account the connection of the mobile to a public or private IPv4 network. This arrangement makes it possible to avoid unrolling the two MIPv4 and MIPv6 mechanisms when the mobile has a dual IPv4 and IPv6 stack.

Several types of tunnel can be built between the mobile and the PGW entity that hosts the home agent functions:

– an IPv6 packet can be encapsulated by an IPv6 header;

– an IPv6 packet can be encapsulated by an IPv4 header. When the mobile is connected to an IPv4 private network, the tunnel must insert a UDP header between the IPv6 and IPv4 headers, for traversal of the NAT (Network Address Translation) device;

– an IPv4 packet can be encapsulated by an IPv6 header;

– an IPv4 packet can be encapsulated by an IPv4 header. When the mobile is connected to an IPv4 private network, the tunnel must insert a UDP header between the two IPv4 headers for traversal of the NAT device.

The direct transfer between the mobile node and the correspondent node is not allowed, the mobile traffic in any case to be controlled by the PGW entity.

9.3. Application to the 4G mobile network

9.3.1. *Trusted Wi-Fi access*

The establishment of the S2c tunnel constitutes one of the different phases of the mobile attachment described in Figure 9.8.

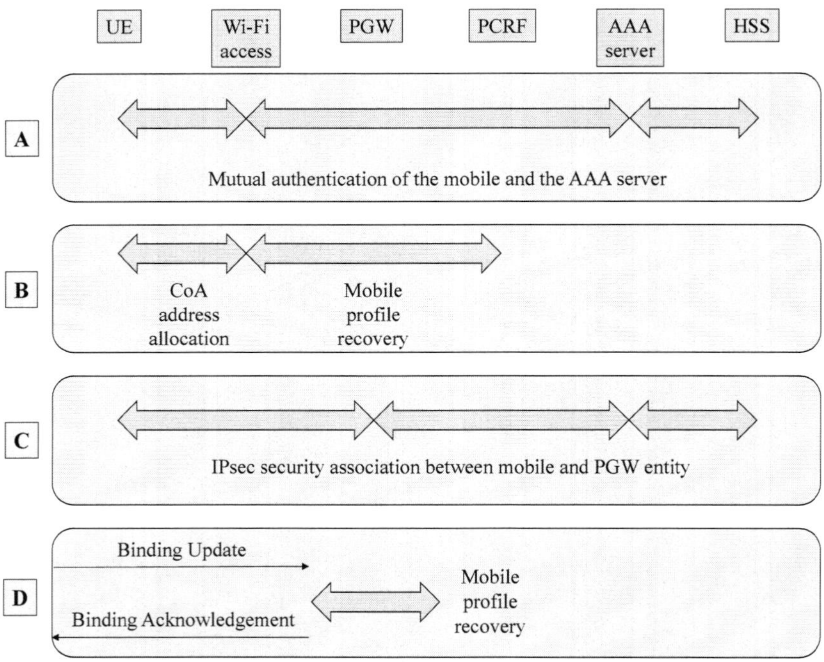

Figure 9.8. *S2c tunnel establishment Trusted Wi-Fi access*

Phase (A) corresponds to the mutual authentication procedure described in Chapter 6. At the end of phase (A), trusted Wi-Fi access has recovered the service profile of the mobile stored in the HSS entity.

Phase (B) corresponds to the configuration of the mobile via trusted Wi-Fi. At the end of phase (B), the mobile recovers its CoA. Trusted Wi-Fi access can also initiate a session with the PCRF to retrieve the profile of the mobile stored in the SPR database.

Phase (C) is the establishment of an IPSec association between the mobile and the PGW entity to protect the DSMIPv6 control messages. The principles for establishing a security association are described in Chapter 7. At the end of phase (C), the PGW entity assigned the mobile its HoA and retrieved the service profile of the mobile stored in the HSS entity.

During phase (D), the mobile communicates to the PGW entity the HoA and CoA in the Binding Update message of the Mobility extension of the IPv6 header. During phase (D), the PGW entity can also initiate a session with the PCRF entity to retrieve the profile of the mobile stored in the SPR entity. The PGW terminates phase (D) by issuing the Binding Acknowledgment message of the Mobility extension of the IPv6 header. At the end of phase (D), the IP tunnel S2c is established between the mobile and the PGW entity.

9.3.2. *Untrusted Wi-Fi access*

The establishment of the S2c tunnel constitutes one of the different phases of the mobile attachment described in Figure 9.9.

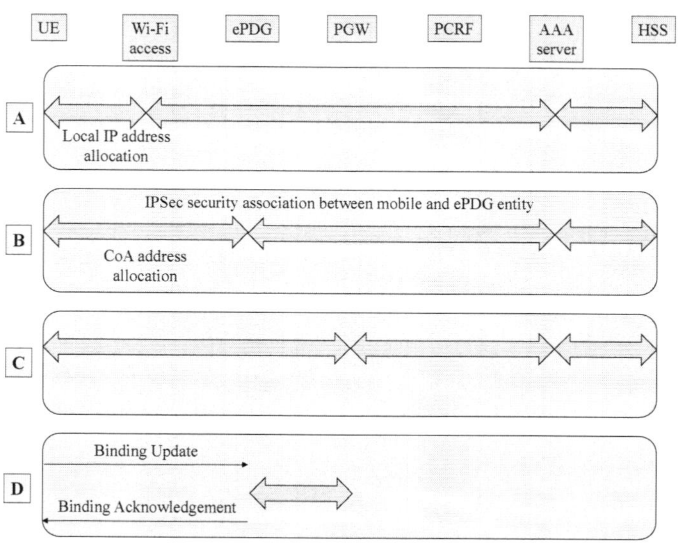

Figure 9.9. *S2c tunnel establishment Untrusted Wi-Fi access*

Phase (A) corresponds to the authentication procedure described in Chapter 6. At the end of phase (A), the untrusted Wi-Fi access has recovered the service profile of the mobile stored in the HSS entity. Untrusted Wi-Fi access provides the mobile with a Local IP Address to start Phase (B) of the procedure.

Phase (B) corresponds to the procedure for establishing the SWu tunnel described in Chapter 7. At the end of phase (B), an IPSec tunnel is established between the mobile and the ePDG entity, the ePDG entity has retrieved the service profile of the mobile stored in the HSS entity and assigned the mobile its CoA.

Phase (C) is the establishment of an IPSec association between the mobile and the PGW entity to protect the DSMIPv6 control messages. At the end of phase (C), the PGW entity allocated the mobile its HoA and retrieved the mobile service profile stored in the HSS entity.

During phase (D), the mobile communicates to the PGW entity the HoA and CoA in the Binding Update message of the Mobility extension of the IPv6 header. During phase (D), the PGW entity can also initiate a session with the PCRF entity to retrieve the profile of the mobile stored in the SPR entity. The PGW entity terminates phase (D) by issuing the Binding Acknowledgment message of the Mobility extension of the IPv6 header. At the end of phase (D), the IP tunnel S2c is established between the mobile and the PGW entity.

9.3.3. *IFOM function*

The IFOM (IP Flow Mobility) function allows the mobile to connect to both LTE (Long-Term Evolution) access and Wi-Fi access simultaneously and to establish multiple sessions from a single connection to the PGW entity.

The IFOM feature also allows mobility of the IP stream, with IP streams belonging to the same application or different applications moving seamlessly between LTE access and Wi-Fi access.

The IFOM function is used to indicate how the IP streams are routed through the different radio access networks and to selectively unload certain traffic (e.g. Internet traffic) to Wi-Fi access while using the LTE access for other traffics (e.g. voice).

The IFOM function requires an evolution of the DSMIPv6 mechanism:

– the mobile can register several CoA associated with the HoA;

– to register the different CoA/HoA correspondences, the mobile generates a BID (Binding Identifier) for each CoA. The mobile requests the registration of its CoA by sending the Binding Update message;

– when the home agent receives the Binding Update message, it copies the BID in its correspondence table (Table 9.1);

– to route the IP flows through a specific access, the mobile must ask the home agent to store traffic selection filters for this access. The mobile includes the Flow Identifier (FID) in the Binding Update message (Table 9.2);

– the mobile assigns a priority level to each BID. If incoming traffic does not match the traffic selection criteria, then the CoA corresponding to the lowest priority will be used.

HOA	CoA	BID
HoA	CoA1	BID1
HoA	CoA2	BID2

Table 9.1. *Correspondence table between the HoA and CoA addresses*

BID	FID	Traffic selection
BID1	FID1	IP source/destination address source/destination port number transport protocol
BID2	FID2	IP source/destination address source/destination port number transport protocol

Table 9.2. *Correspondence table between the BID and FID identifiers*

10

Network Discovery and Selection

10.1. Mechanisms defined by 3GPP organization

10.1.1. *ANDSF function*

The selection of the access network and the management of the traffic between LTE (Long-Term Evolution) access and Wi-Fi (Wireless Fidelity) access are supported by the ANDSF (Access Network Discovery and Selection Function) server.

The information provided by the ANDSF server has a tree structure of management objects (MO) that use an extensible markup language (XML).

The mobile can access the ANDSF server via Wi-Fi access to the Internet or via Wi-Fi access or LTE access to the 4G mobile network.

The ANDSF server can push the information to the mobile (push mode) or the mobile can interrogate the ANDSF server and receive the corresponding information (pull mode). If the mobile submits a request, then it may also include other information in its request, such as its location and discovered radio access networks.

The mobile can discover the ANDSF server in one of the following three ways:

– static configuration;

– DNS (Domain Name Service) resolution, for which a specific full qualified domain name (FQDN) is used:

andsf.mnc <MNC> .mcc <MCC> .pub.3gppnetwork.org

– dynamic configuration by a DHCP (Dynamic Host Configuration Protocol) server.

The ANDSF server determines the access on which the mobile must transfer the IP (Internet Protocol) flow in the following cases:

– the mobile is able to route IP packets via a single type of access, LTE or Wi-Fi;

– the mobile is able to route different IP packets for the same PDN (Packet Data Network) connection via different access networks;

– the mobile is able to route IP packets for different PDN connections via different access networks.

The information provided by the ANDSF server may also be preconfigured by the home operator on the terminal or provisioned on the universal integrated circuit card (UICC).

10.1.1.1. *ANDI*

Following a mobile request, the ANDSF server can provide access network discovery information (ANDI) in the vicinity of the mobile (Figure 10.1):

– the types of access technologies, such as the Wi-Fi interface;

– the identifier of the access network, such as the service set identifier (SSID);

– specific information on the characteristics of the radio interface, such as the frequency of the radio channel;

– the conditions indicating when the ANDI is valid.

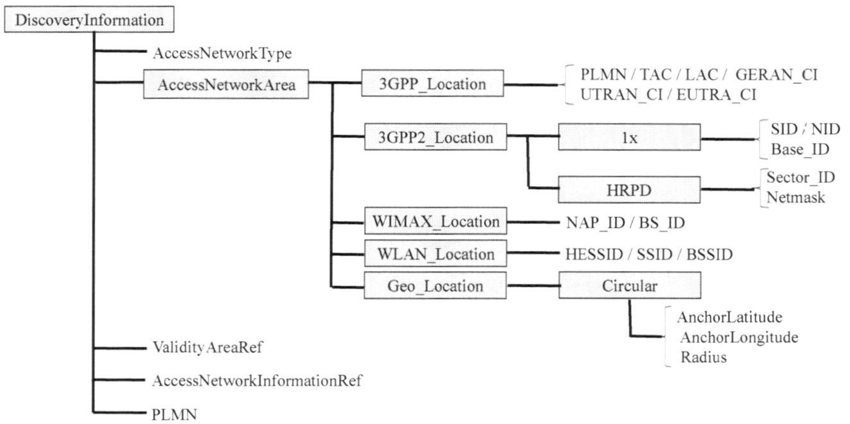

Figure 10.1. *ANDI information*

10.1.1.2. *ISMP*

The inter-system mobility policy (ISMP) provides the mobile with the rules for routing IP packets over LTE or Wi-Fi interfaces. The mobile uses these rules when it cannot access both interfaces simultaneously.

Figure 10.2 describes the structure of the managed objects (MO) for the ISMP.

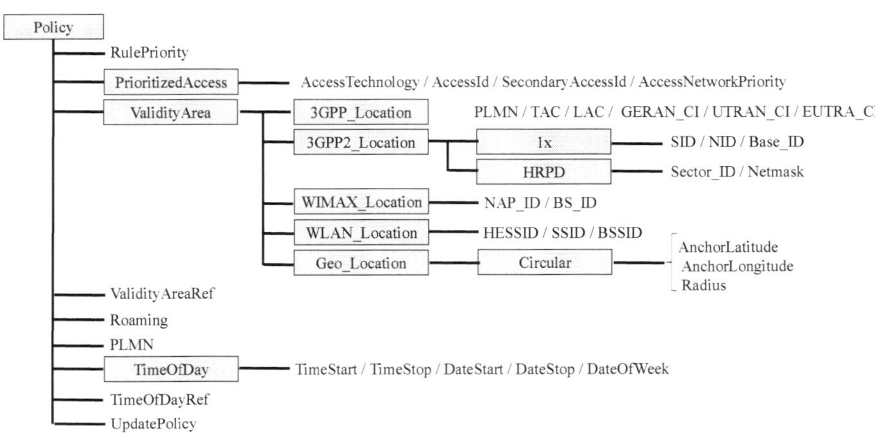

Figure 10.2. *ISMP policy*

Each ISMP rule includes the following information:

– the conditions that indicate when the rule is valid. These conditions may include, for example, a duration or a location area;

– a priority list of access technologies that indicates the order in which they are preferred or restricted for the connectivity to the evolved packet core (EPC);

– a rule priority that indicates the priority of this rule over other ISMP rules provided by the same 4G mobile network.

10.1.1.3. *ISRP*

The inter-system routing policy (ISRP) is a set of operator-defined rules that determine how the mobile should route traffic across multiple radio access interfaces.

The IFOM (IP Flow Mobility) rules identify a prioritized list of radio access technologies that should be used by the mobile to route the different IP packets of a PDN connection that corresponds to an access point name (APN).

As the PDN connection is anchored in the EPC network, mobility for each IP flow is provided between LTE and Wi-Fi accesses.

Figure 10.3 describes the structure of managed objects (MO) for IFOM rules.

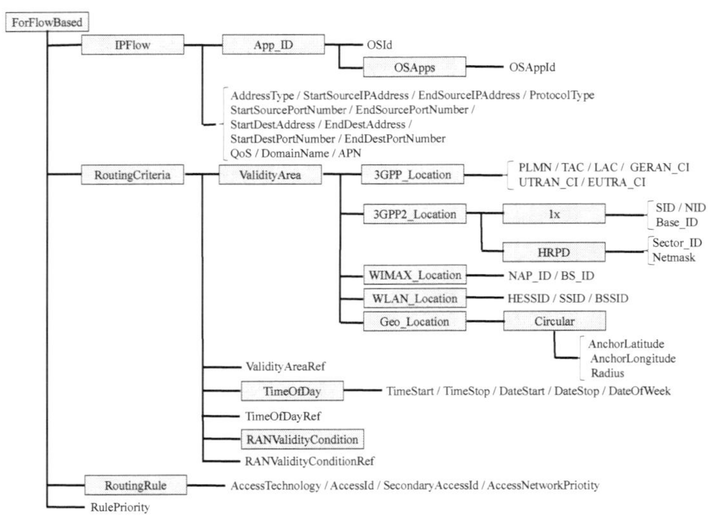

Figure 10.3. *IFOM rules*

An IFOM rule can also identify which radio accesses are restricted for the traffic, such as the prohibition to use the Wi-Fi interface for IP packets containing voice.

Each IFOM rule can identify the traffic based on the IP address of the source or destination, the transport protocol, the port numbers of the source or destination, the DSCP (DiffServ Code Point) or TC (Traffic Class) field.

The MAPCON (Multiple-Access PDN Connectivity) rules identify a prioritized list of radio access technologies that should be used by the mobile to route each PDN connection that corresponds to an access point name (APN).

As the PDN connection is anchored in the EPC network, mobility for each PDN connection is provided between LTE and Wi-Fi accesses.

Figure 10.4 describes the structure of managed objects (MO) for MAPCON rules.

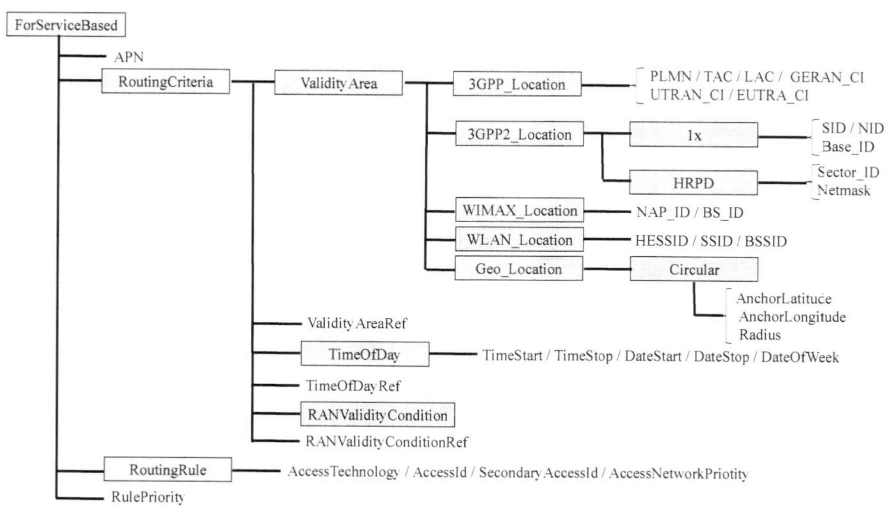

Figure 10.4. *MAPCOM rules*

A MAPCON rule can also identify which radio access is restricted for PDN connections, for example, prohibiting the use of the Wi-Fi interface for certain types of access points (APN).

The NSWO (Non-Seamless WLAN Offload) rules identify which IP packets should be offloaded by Wi-Fi access to the Internet network without crossing the EPC network.

Because streams are not anchored in the EPC network, mobility for each IP stream is not assured between LTE and Wi-Fi accesses.

It is possible to restrict or allow the offloading of traffic to specific Wi-Fi access networks.

Figure 10.5 describes the structure of managed objects (MO) for NSWO rules.

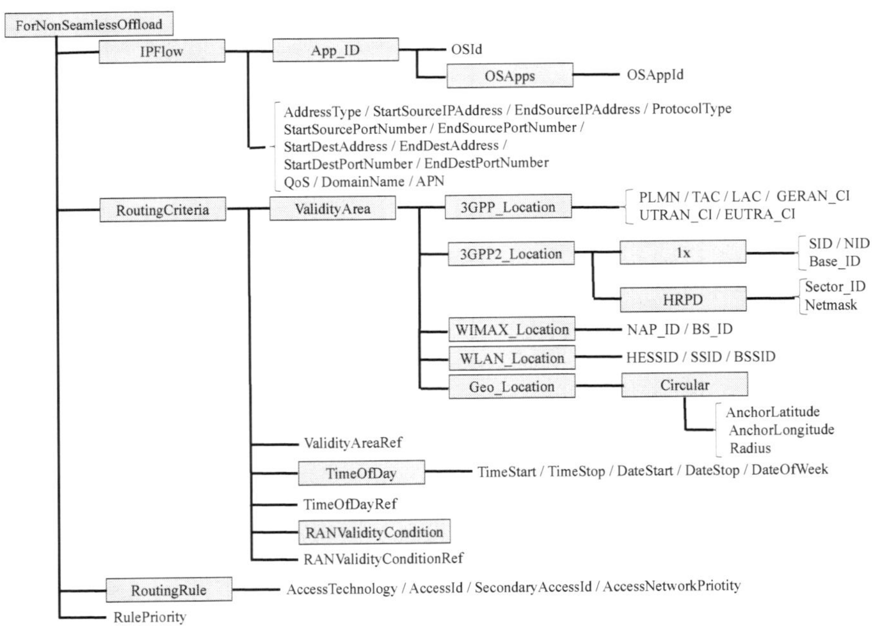

Figure 10.5. *NSWO rules*

10.1.1.4. *IARP*

The IARP (Inter-APN Routing Policy) rules determine which traffic should be routed across different PDN connections and which traffic should be offloaded by Wi-Fi access to the Internet network (Figure 10.6).

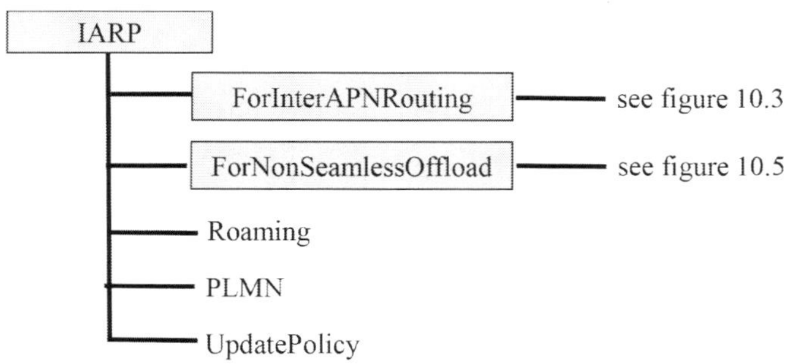

Figure 10.6. *IARP rules*

The rules for the access point (APN) identify a prioritized list of access point names that should be used by the mobile to route traffic that matches IP traffic filters.

The rules for NSWO identify which traffic should be offloaded for Wi-Fi access to the Internet network.

When the mobile has both the IARP rule and the ISRP rule simultaneously, it first evaluates the IARP rule to determine how to route an IP stream. If the IP stream does not match any IARP rules, the mobile evaluates the active ISRP rules to determine how to route the IP stream.

10.1.1.5. *WLANSP*

The WLANSP is a set of rules that determine how the mobile selects a Wi-Fi access network.

Figure 10.7 describes the structure of the managed objects (MO) for the WLANSP.

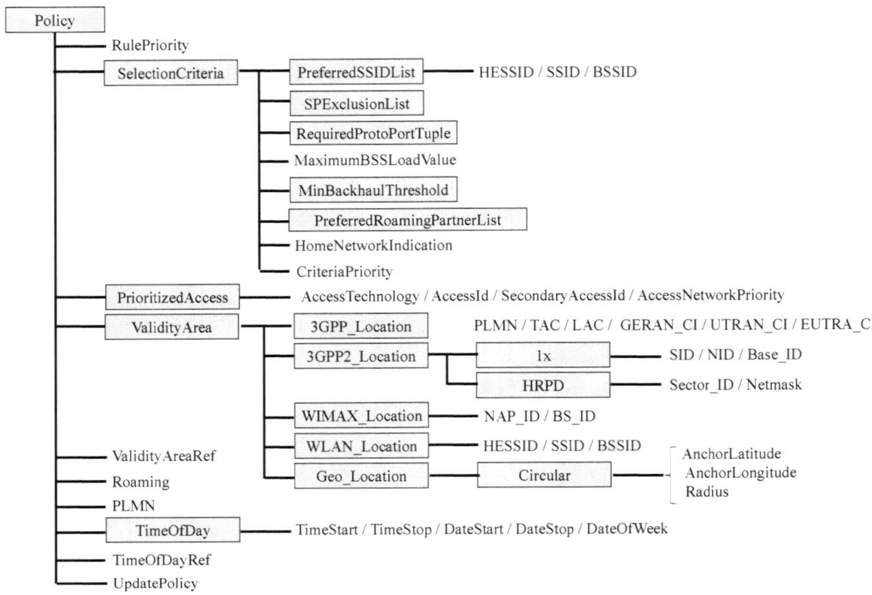

Figure 10.7. *WLANSP policy*

Each WLANSP rule includes the following information:

– the conditions indicating when the rule is valid. The conditions of validity can include time, geolocation and location of the network, such as the location area;

– the selection criteria that must be fulfilled by the Wi-Fi access network to be eligible, such as cell load or transmission network throughput.

10.1.1.6. *Wi-Fi access network preferences*

Network preferences include information that helps the mobile to select a Wi-Fi access network.

The network preferences indicate whether the network prefers the mobile to establish a PDN connection using the S2a architecture.

In the case of a PDN connection using the S2b architecture, the network preferences indicate the identity of the evolved packet data gateway (ePDG).

The EHSP (Equivalent Home Service Providers) information contains a list of service providers that are equivalent to the mobile home network. Each service provider is identified with a domain name.

The PSPL (Preferred Service Provider List) information contains a preferred list of service providers.

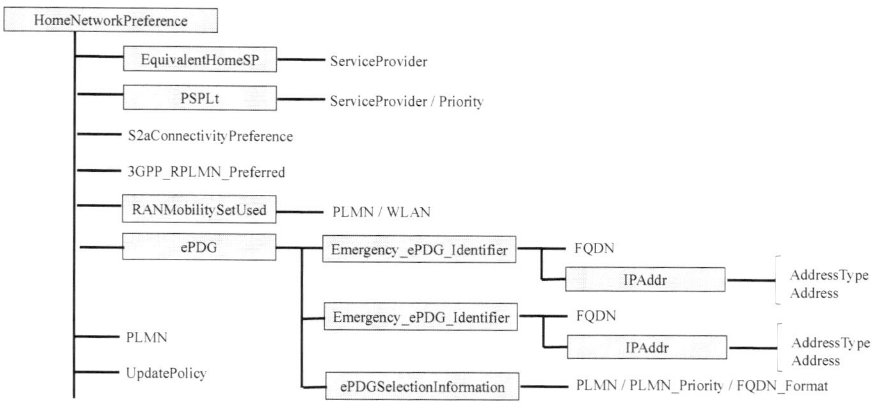

Figure 10.8. *Wi-Fi access network preferences*

10.1.2. *RAN assistance*

The evolved node B station (eNB) can provide mobile support information. This information includes the following parameters:

– the thresholds for access to the LTE interface;

– the thresholds for the Wi-Fi interface;

– the offload preference indication (OPI).

The thresholds for the LTE interface define the high and low values of the radio parameters, for example, the average value of the preference signal received power (RSRP).

The thresholds for the Wi-Fi interface define the high and low values of the access parameters, such as the received signal strength indication (RSSI) of the beacon, the transmission network throughput and the load of the radio channel.

The OPI is a one-dimensional bitmap that can be used by mobiles to determine when they should move certain traffic to Wi-Fi access or LTE access. The meaning of each bit is operator specific.

The thresholds and parameters can affect the validity of ANDSF rules and thus make these rules subject to the conditions defined by the eNB entity.

The thresholds and parameters can be used by the following ANDSF rules:

– the ISRP rules, including IFOM rules, MAPCON rules and NSWO rules;

– the IARP rules, including rules related to the access point (APN) rules and NSWO.

The selection of the Wi-Fi interface and the routing behavior for the mobile must be controlled either by the ANDSF rules or by the rules provided by the eNB entity, and not by a combination thereof.

The only exception is the simultaneous enforcement of the rules provided by the eNB entity and the IARP rules for the access point (APN).

10.2. Mechanisms defined by IEEE and WFA organizations

Before associating with an access point, the mobile requires information on the services provided by the Wi-Fi access networks, from GAS (Generic Advertisement Service) frames that are Action-type management frames.

The Public Action field, in the byte immediately after the Category field, differentiates the types of Action frames.

GAS frames provide transparent transport of a list with ANQP (Access Network Query Protocol) elements to communicate information.

The Interworking element in the management frames Beacon or Probe Response indicates that the GAS protocol is supported (Figure 10.9).

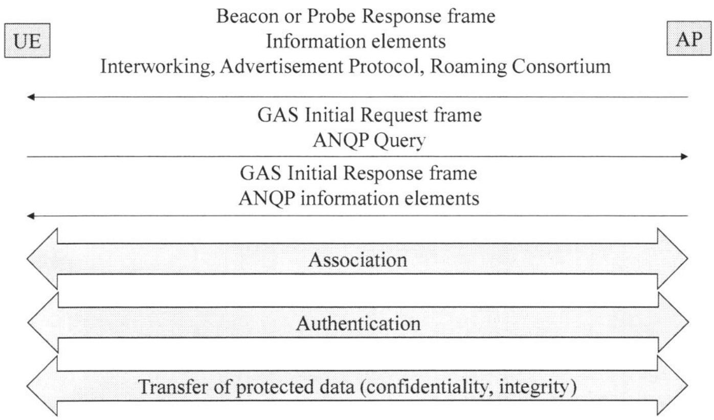

Figure 10.9. *GAS/ANQP exchanges*

The Advertisement Protocol element in the management frames Beacon or Probe Response indicates that the ANQP protocol is supported (Figure 10.9).

The mobile transmits a request in a GAS Initial Request frame, and the access point provides the requested information or information on how to receive the response in the GAS Initial Response frame (Figure 10.9).

The response to the GAS Initial Request frame is provided in this case in one or more GAS Comeback Response frames.

The response to the GAS Initial Request frame shall not be shared between an Initial Response GAS frame and one or more Comeback Response GAS frames.

The access point has the information elements or transfers the request from the mobile to an ANQP server.

The IEEE (Institute of Electrical and Electronics Engineers) has defined a subset of ANQP information elements (Table 10.1).

The WFA (Wi-Fi Alliance) organization has completed this list as part of Passpoint certification based on Hotspot 2.0 features (Table 10.1).

ANQP information elements	Specification	Function
3GPP Cellular Network	IEEE	Identification and authentication methods from the service provider
NAI Realm	IEEE	
Roaming Consortium	IEEE	
Domain Name	IEEE	Identification of the Wi-Fi access network
Venue Name	IEEE	
Operator's Friendly Name	WFA	
IP Address Type Availability	IEEE	
WAN Metrics	WFA	Characteristics of the Wi-Fi access network
Connection Capability	WFA	
Operating Class Indication	WFA	
Network Authentication Type	IEEE	
OSU Providers List	WFA	Online registration
Icon Request & Response	WFA	
HS Query List	WFA	Capacity request
HS Capability List	WFA	
NAI Home Realm Query	WFA	

Table 10.1. *ANQP information elements*

10.2.1. *Information elements provided by the beacon*

10.2.1.1. *HESSID element*

If two access points have different SSIDs, then they are considered as different Wi-Fi networks. If two access points have the same SSID, then they are considered as part of the same wireless network.

However, SSIDs are not globally administered, and it is possible that two access points with the same SSID are actually different Wi-Fi networks.

The homogeneous extended service set identifier (HESSID) allows mobiles to detect this condition. When two access points of two different Wi-Fi networks have the same SSID, the two networks are differentiated by two different HESSIDs.

The HESSID is included in the Interworking element in Beacon or Probe Response frames.

The HESSID is a MAC (Medium Access Control) address. The HESSID value has the same value as the basic service set identifier (BSSID) of one of the access points.

10.2.1.2. *Access Network Type field*

The Access Network Type field is included in the Interworking element. Mobiles can use this information when selecting an access point.

The Access Network Type field indicates the type of network to which the access point is connected: pay public network, free public network, private network and private network with guest access.

10.2.1.3. *Internet Available field*

The Internet Available field is included in the Interworking element. This field informs mobiles if access to the Internet is available at the access point, which may not be the case in environments where the operator (e.g. a museum) may limit Wi-Fi access to only local content.

10.2.1.4. *BSS Load element*

The BSS Load information element contains information on the use of the radio channels and the number of associated mobiles on the access point. The mobile uses this information when selecting a network.

10.2.2. *Information elements provided by the ANQP server*

10.2.2.1. *3GPP Cellular Network element*

The information element 3GPP Cellular Network contains the identity of the 4G mobile network. It allows the mobile to check from its universal subscriber identity module (USIM) if the Wi-Fi network operator has a roaming agreement with the 4G mobile network operator.

The identity of the 4G mobile network consists of the mobile country code (MCC) and the mobile network code (MNC) allocated to the operator.

If the information element 3GPP Cellular Network matches any identity stored in the mobile, then it prioritizes this access point for the association.

10.2.2.2. *NAI Realm element*

The information element NAI Realm provides a list of domains identified by the network access identifier (NAI) for service providers that can authenticate a mobile with either a user ID or a password or a certificate.

Each entry in the NAI Realm list can identify the EAP (Extensible Authentication Protocol) methods that are supported for authentication.

10.2.2.3. *Roaming Consortium element*

The information element Roaming Consortium provides a list of roaming consortium identifiers and service provider partners with roaming agreements.

The information element Roaming Consortium is broadcasted in the management frame Beacon or transmitted in the Probe Response frame. A mobile may request an information element Roaming Consortium if the information received is insufficient for the selection of the network.

10.2.2.4. *Domain Name element*

The information element Domain Name provides a list of one or more domain names of the entity that operates the Wi-Fi network.

The mobile uses the domain name to determine whether access to this Wi-Fi network through this access point is considered access to its home network or to a visited network.

10.2.2.5. *Venue Name element*

The information element Venue Name provides venue names that can be used to help the mobile to select the access point. The names of the venue can be included in the same language or in different languages.

10.2.2.6. *Operator's Friendly Name element*

The information element Operator's Friendly Name provides the friendly name of the Wi-Fi network operator.

The mobile can obtain the name of the operator via GAS/ANQP requests to help the user when manually selecting access points.

10.2.2.7. *IP Address Type Availability element*

The information element IP Address Type Availability provides information about IP addresses and port numbers:

– Wi-Fi access point allocates a public IPv4 address;

– Wi-Fi access point allocates a private IPv4 address;

– the combination of the Wi-Fi access network and the core network allocates a dual NAT IPv4 address;

– Wi-Fi access point allocates an IPv6 address.

10.2.2.8. *WAN Metrics element*

The information element WAN Metrics provides information about the link that connects the access point to the Internet network: the state of the link, the value of the bit rates for each direction of transmission.

The access point may also provide additional information, such as the load for each direction of transmission.

The mobile uses this information to make network selection decisions. The mobile determines whether the available rate level is compatible with the application need.

10.2.2.9. *Connection Capability element*

The information element Connection Capability provides information about the allowed values of the Protocol field of the IPv4 header or Next Header field of the IPv6 header, and port numbers.

The mobile uses this information to make network selection decisions. The mobile determines whether the allowed values are compatible with the characteristics of the application.

10.2.2.10. *Operating Class Indication element*

The information element Operating Class Indication provides information about the radio channels and frequency bands used by the access point.

The mobile uses this information to make network selection decisions. If the mobile supports the 2.4 and 5 GHz frequency bands, and if these two frequency bands are available at the access point, then the mobile will select the 5 GHz band.

10.2.2.11. *Network Authentication Type element*

The information element Network Authentication Type provides a list of authentication types:

– the network requires the user to accept the terms and conditions;

– the network supports online registration;

– the network infrastructure performs HTTP/HTTPS redirection;

– the network supports a DNS redirection.

10.2.2.12. *OSU Providers List element*

The information element OSU Providers List contains a list of entities that offer an online registration service.

The information element OSU Providers List provides a list of available icons that can be displayed by the mobile. This list contains the definition of the image, the image type, the language and the name of the file. This information allows the mobile to determine the icon to download and the file name of the icon to recover.

10.2.2.13. *Icon Request & Response element*

The information element Icon Request & Response allows the mobile to send a request containing the file name of the icon and allows the access point to return the answer containing the download status code, the length of the type of icon, data length and binary icon data.

10.2.2.14. *HS Query List element*

The information element HS Query List is transmitted by the mobile to obtain information simultaneously on several elements of ANQP information.

The information element HS Query List is transmitted in a GAS Query Request frame.

10.2.2.15. *HS Capability List element*

The information element HS Capability List tells the mobile which ANQP elements are supported by the access point.

The information element HS Capability List is transmitted in a GAS Query Response frame.

10.2.2.16. *NAI Home Realm Query element*

The information element NAI Home Realm Query enables the mobile to determine whether the domains for which it has security information correspond to the service providers whose networks are accessible at the access point.

11

Carrier Aggregation

11.1. Functional architecture

The integration of Wi-Fi technology into the 4G mobile network results in the sharing of sessions between LTE (Long-Term Evolution) access (e.g. VoLTE session) and Wi-Fi access (e.g. Internet session). The aggregation of LTE and Wi-Fi access is done at the PDN gateway (PDN). The consideration of Wi-Fi access impacts the architecture of the evolved packet core (EPC).

The aggregation of the carriers results in a sharing of the IP (Internet Protocol) packets between the different accesses. For the downstream direction (respectively the upstream direction), the sharing is performed by the evolved node B station (eNB) (respectively the mobile) and the reassembly is provided by the mobile (respectively the entity eNB). This operation is performed exclusively in the enhanced universal terrestrial radio access network (E-UTRAN) and does not impact the EPC network (Figure 11.1).

LTE and Wi-Fi carrier aggregation can be implemented with collocated or remote eNB and AP (Access Point) entities. The Xw interface is the point of reference between the eNB and AP entities when they are distant (Figure 11.1).

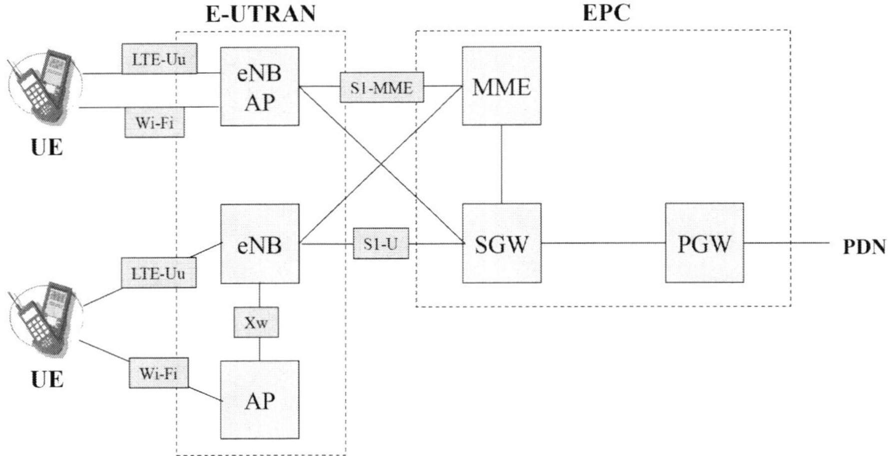

Figure 11.1. *Functional architecture for LTE and Wi-Fi carrier aggregation*

The eNB entity is the anchor point for the data exchanged with the mobile, belonging to the user plane (the IP packets) and the control plane and connects to the EPC network:

– at the level of the mobility management entity (MME), via the S1-MME interface, for the S1-AP signaling;

– at the level of the serving gateway (SGW), via the S1-U interface, for the S1bearer.

11.2. Protocol architecture

11.2.1. *LWA*

At the radio interface LTE-Uu, between the mobile and the eNB entity, the traffic data correspond to IP packets and the signaling data relate to RRC (Radio Resource Control) messages exchanged between the mobile and the eNB entity and NAS (Non-Access Stratum) messages exchanged between the mobile and the MME entity.

The traffic and signaling data is encapsulated by the data link layer, broken down into three sub-layers:

– packet data convergence protocol (PDCP);

– radio link control (RLC);

– medium access control (MAC).

LWA (LTE/WLAN Aggregation) occurs at the PDCP layer. The entity eNB carries out a switching of the bearers between, on the one hand, the S1 bearers and, on the other hand (Figures 11.2 and 11.3):

– the LTE bearer, for which the data transits only on the LTE access;

– the shared LWA bearer, for which the data can pass on both LTE and Wi-Fi accesses;

– the switched LWA bearer, for which the data only passes over the Wi-Fi access.

The LWA bearer is controlled by the eNB entity from measurement reports transmitted by the mobile.

The PDCP frames transmitted over the Wi-Fi access are encapsulated by an LWAAP (LWA Adaptation Protocol) header containing the logical channel identifier (LCID) of the radio bearer.

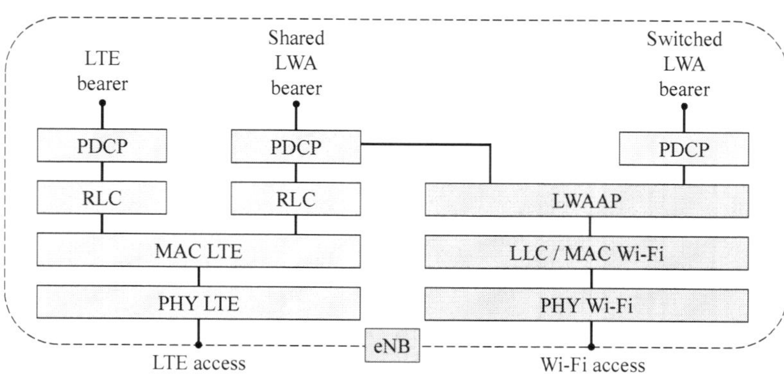

Figure 11.2. *Protocol architecture for LWA aggregation eNB and AP entities are collocated*

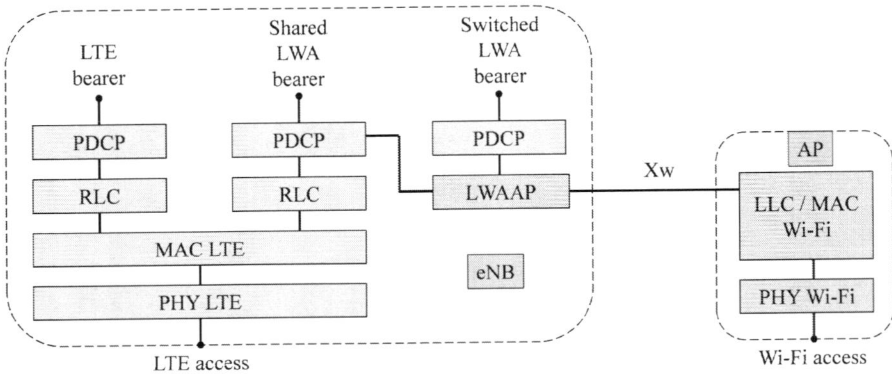

Figure 11.3. *Protocol architecture for LWA aggregation eNB and AP entities are distant*

On LTE access, the LCID is carried by the MAC layer. The recipient uses the LCID to reassemble the PDCP frames of the same bearer.

The re-sequencing of the PDCP frames received by the two LTE and Wi-Fi accesses is performed by the PDCP.

Frames transported on an LWA bearer are only those acknowledged and those corresponding to RLC frames using acknowledgment mode (AM) on the LTE interface.

The Type field of the LLC header for Wi-Fi access is set to hexadecimal 9E65. The mobile uses this value to determine that the frame comes from an LWA bearer.

When the eNB and AP entities are distant, the eNB entity can be connected to multiple AP entities via the Xw interface that supports the traffic and control data (Figure 11.3).

The NAS signaling data is carried on the S1-MME interface, between the MME and eNB entities, and then on the Xw-C (Control) interface, between the eNB and AP entities.

Traffic data, corresponding to the IP stream, is transported in a GTP-U (GPRS Tunneling Protocol User) tunnel:

– on the S1-U interface, between SGW and eNB entities;

– on the Xw-U (User) interface, between the eNB and AP entities.

Mutual authentication is based on the EAP-AKA (Authentication and Key Agreement) method described in Chapter 6.

During the procedure of attaching the mobile to the 4G mobile network, the home subscriber server (HSS) retrieves the Ki key allocated to the mobile, draws a random (RAND) and calculates the integrity key (IK) and cipher key (CK).

The Ki key is a secret key, shared between the HSS entity and the mobile, created during the subscription.

The HSS entity computes the K_{ASME} key from the IK and CK and passes the K_{ASME} key and the random to the mobility management entity (MME).

From the K_{ASME} key, the MME entity calculates the Ke_{NB} key and transmits it to the eNB entity.

The MME entity transmits the random (RAND) to the mobile that performs the same operations to generate the Ke_{NB} key.

The pairwise master key (PMK) used for the four-way handshake procedure is the S-K_{WT} key derived from the Ke_{NB} key.

11.2.2. *LWIP aggregation*

The LWIP (LTE/WLAN radio level integration with IPsec tunnel) aggregation only applies to IP packets of the S1 bearer. The RRC (Radio Resource Control) and signaling messages, which are exchanged between the mobile and the eNB entity, are carried on the LTE interface (Figure 11.4).

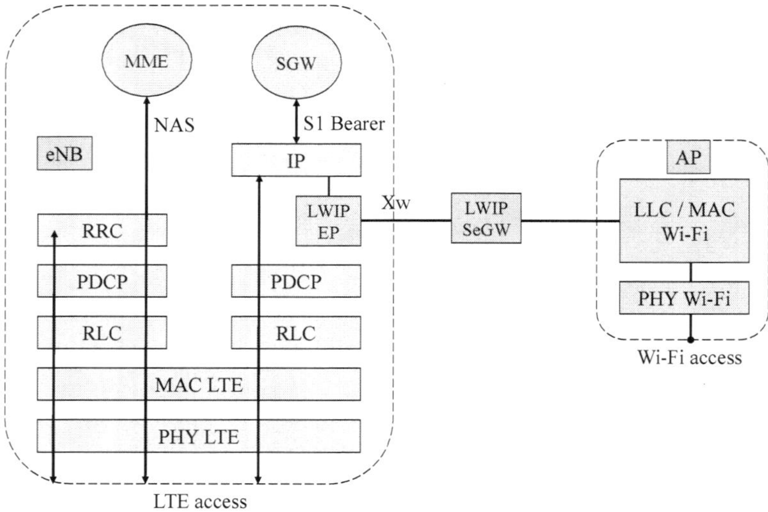

Figure 11.4. *Protocol architecture for LWIP aggregation*

IP packets are transported between the eNB entity and the mobile in the LWIP tunnel. The LWIPEP (LWIP Encapsulation Protocol) header contains the LCID of the radio bearer.

The LWIP tunnel is protected between the mobile and the security gateway (SeGW) through an IP Security (IPSec) mechanism. Only one IPSec mechanism is mounted for all LWIP tunnels.

The LWIP tunnel is transmitted in a GTP-U tunnel on the Xw interface, between the eNB and the SeGW entities.

The IKE procedure for the IPSec mechanism is initialized after the association of the mobile to the Wi-Fi access point and authentication based on the EAP-AKA method.

Each bearer is configured so that the downstream direction or the upstream direction or both directions of transmission pass through the tunnel protected by the IPSec mechanism.

For the downstream, IP packets are transmitted either on the LTE interface only, or on the Wi-Fi interface only, or simultaneously on both

LTE and Wi-Fi interfaces. In the latter case, the mobile can receive IP packets not in sequence.

For the upstream, IP packets are transmitted either on the LTE interface only, or on the Wi-Fi interface only.

11.2.3. *LAA aggregation*

LAA (Licensed Assisted Access) aggregation consists of using the 5-GHz U-NII (Unlicensed-National Information Infrastructure) band to transmit a 3GPP compliant LTE signal.

The radio channel operating in the licensed band, used as the primary channel, supports control plane (signaling) and traffic plane (IP packets).

The radio channel operating in the U-NII band, used as a secondary channel, supports only the data of the traffic plane.

11.3. Procedures

11.3.1. *LWA*

11.3.1.1. *WT Addition procedure*

The WT Addition procedure is initialized by the eNB entity and is used to establish the mobile context at the AP to provide mobile resources over the Wi-Fi interface (Figure 11.5).

1) The eNB entity transmits to the access point (AP) the message Xw-AP WT Addition Request in order to allocate resources to the mobile, indicating the characteristics of the LWA bearer.

2) If the access point can accept the resource request, it responds with the message Xw-AP WT Addition Request Acknowledge.

3) The eNB entity sends the message RRC *ConnectionReconfiguration* to the mobile, indicating the configuration of the radio resource.

4) The mobile applies the new configuration and responds to the eNB entity with the message RRC *ConnectionReconfigurationComplete*.

5) The mobile associates with the access point, which then transmits the message Xw-AP WT Association Confirmation to the eNB entity.

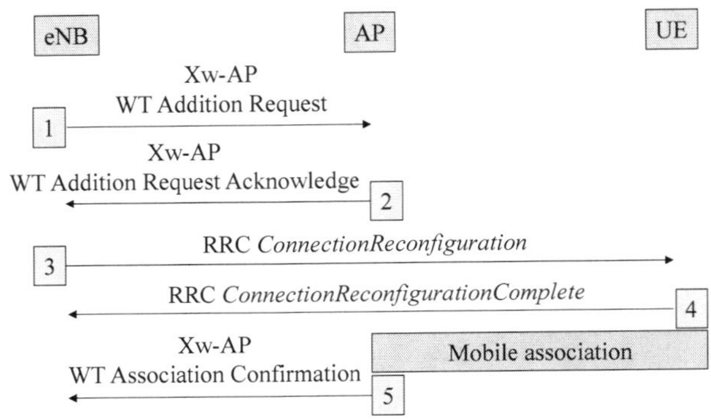

Figure 11.5. *WT Addition procedure*

11.3.1.2. *WT Modification procedure*

The WT Modification procedure can be initialized either by the eNB entity or by the access point and can be used to modify, set or release bearer contexts or to modify other properties of the mobile context.

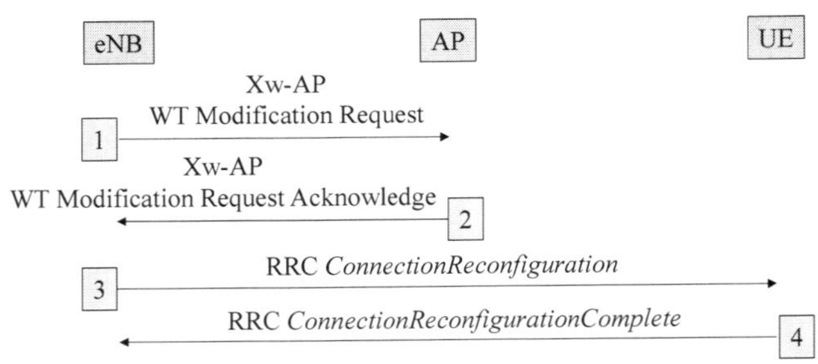

Figure 11.6. *WT Modification procedure initiated by the eNB entity*

The WT Modification procedure initiated by the eNB entity is described in Figure 11.6.

1) The eNB entity sends the message Xw-AP WT Modification Request to request the AP to modify the specific bearer resources.

2) If the access point accepts the request, it applies the configuration modification to the resource and responds with the message Xw-AP WT Modification Request Acknowledge.

3) If the modification requires a new configuration for the mobile, the eNB entity sends the message RRC *ConnectionReconfiguration*, including the new configuration of the Wi-Fi radio resource.

4) The mobile applies the new configuration and responds with the message RRC *ConnectionReconfigurationComplete*.

The WT Modification procedure initiated by the access point is described in Figure 11.7.

1) The access point sends the message Xw-AP WT Modification Required to the eNB entity to modify the radio resources of the Wi-Fi access.

2) The eNB responds with the message Xw-AP WT Change Confirm.

3) If the modification requires a new configuration for the mobile, the eNB entity sends the message RRC *ConnectionReconfiguration*, including the new configuration of the Wi-Fi radio resource.

4) The mobile applies the new configuration and responds with the message RRC *ConnectionReconfigurationComplete*.

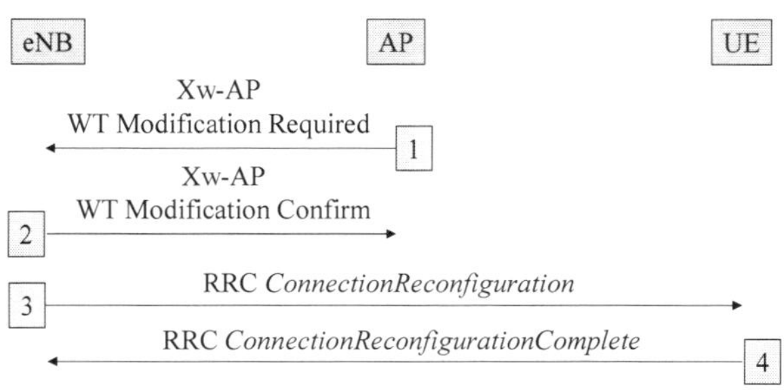

Figure 11.7. *WT Modification procedure initiated by the access point*

11.3.1.3. *WT Release procedure*

The WT Release procedure can be initialized either by the NB entity or by the access point and is used to initiate the release of the mobile context at the access point. The recipient cannot reject the request.

The WT Release procedure initiated by the eNB entity is described in Figure 11.8.

1) The eNB entity sends the message Xw-AP WT Release Request to request the Wi-Fi access point to release the allocated radio resources over the Wi-Fi access.

2) If necessary, the eNB entity sends the message RRC *ConnectionReconfiguration* to the mobile indicating the release of the radio resources.

3) The mobile responds with the message RRC *ConnectionReconfigurationComplete*.

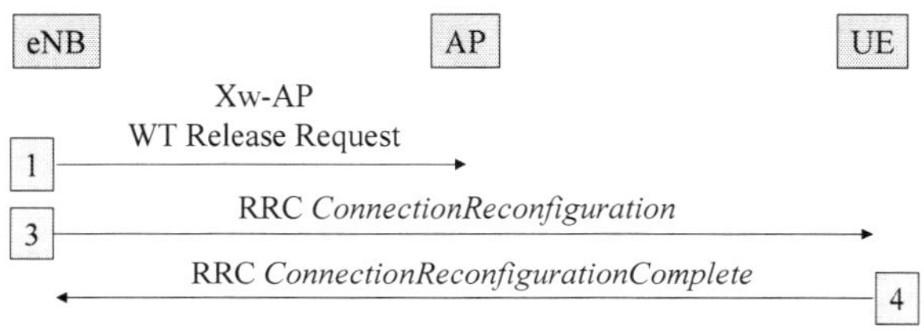

Figure 11.8. *WT Release procedure initiated by the eNB entity*

The WT Release procedure initiated by the access point is described in Figure 11.9.

1) The access point sends the message Xw-AP WT Release Required to the eNB entity to request the release of radio resources from the Wi-Fi access.

2) The eNB entity responds with the message Xw-AP WT Release Confirm.

3) If necessary, the eNB entity sends the message RRC *ConnectionReconfiguration* to the mobile indicating the release of the radio resources.

4) The mobile responds with the message RRC *ConnectionReconfigurationComplete*.

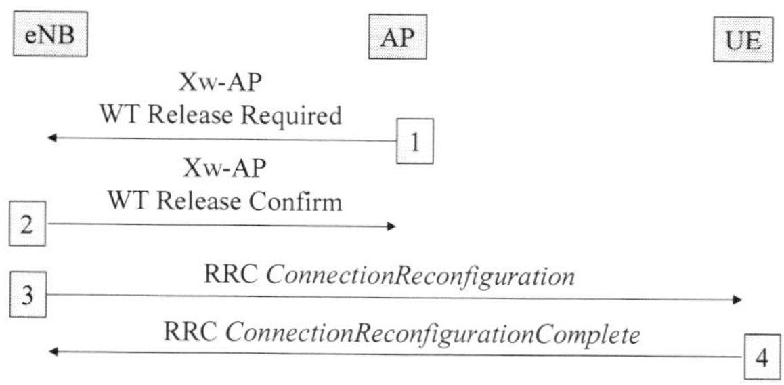

Figure 11.9. *WT Release procedure initiated by the access point*

The procedure for changing the access point is initiated by the eNB entity and used to transfer the mobile context from a source AP to a target AP. This procedure is performed using the WT Release and WT Addition procedures.

11.3.2. *LWIP aggregation*

The procedure for establishing the LWIP and IPSec tunnels is described in Figure 11.10.

1) The eNB entity configures the mobile with the message RRC *ConnectionReconfiguration* to perform measurements on Wi-Fi access in order to start the LWIP and IPSec tunnels establishment.

2) The mobile applies the new configuration and responds with the message RRC*ConnectionReconfigurationComplete.*

3) The mobile sends to the eNB entity the message RRC *WLANMeasurements* containing the measurements performed on the Wi-Fi access.

4) The eNB entity sends the message Xw-AP LWIP Addition Request to request the security gateway (SeGW) to allocate resources for IPSec tunnel establishment.

5) If the security gateway accepts the request, it responds with the message Xw-AP LWIP Addition Request Acknowledge.

6) The eNB entity sends the message RRC *ConnectionReconfiguration* to the mobile to establish the LWIP tunnel.

7) The mobile applies the new configuration and responds with the message RRC *ConnectionReconfigurationComplete.*

8) The mobile sends the confirmation of the association with the Wi-Fi access point to the eNB entity in the message RRC *WLANConnectionStatusReport.*

9) The eNB entity sends the message RRC *ConnectionReconfiguration* to the mobile to establish the IPSec tunnel and can configure the bearers that will use the IPSec tunnel.

10) The mobile applies the new configuration and responds with the message RRC *ConnectionReconfigurationComplete.*

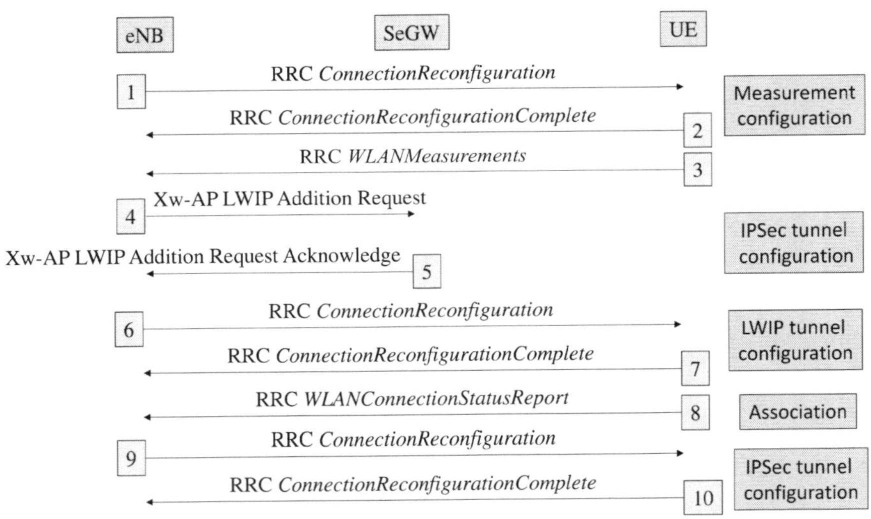

Figure 11.10. *LWIP and IPSec tunnel establishment*

11.3.3. *LAA aggregation*

The access mechanism to the radio channel is different for the LTE and Wi-Fi interfaces.

For the LTE radio interface, access to the radio channel is controlled by the eNB entity.

For the Wi-Fi radio channel, access to the radio channel uses the CSMA/CA (Carrier Sense Multiple Access/Collision Avoidance) mechanism.

The eNB entity or the mobile applies the LBT (Listen Before Talk) mechanism before transmitting in the U-NII radio channel.

The equipment uses energy sensing to determine the presence or absence of other signals on the radio channel during the CCA (Clear Channel Assessment) observation time.

The LBT mechanism has two options: frame-based equipment (FBE) and load-based equipment (LBE).

For the FBE option, the equipment operates on the basis of a synchronization with a fixed frame period.

At the end of the frame period, the equipment performs a CCA check on the radio channel. If the channel is free, then the data is transmitted immediately to the beginning of the next frame. If the channel is busy, then another CCA check is performed at the next frame period (Figure 11.11).

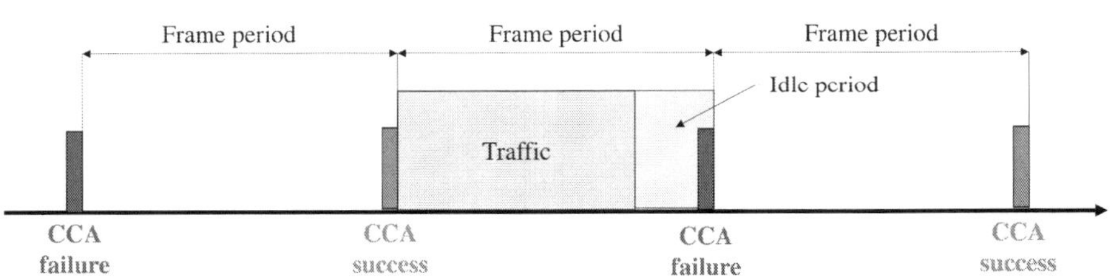

Figure 11.11. *LBT mechanism –FBE option*

For the LBE option, the device performs CCA control whenever there is data to transmit. If the channel is free, then the data is transmitted immediately. If the channel is busy, then the device must wait until the timer for the backoff mechanism expires (Figure 11.12). This timer is decremented when the radio channel is free.

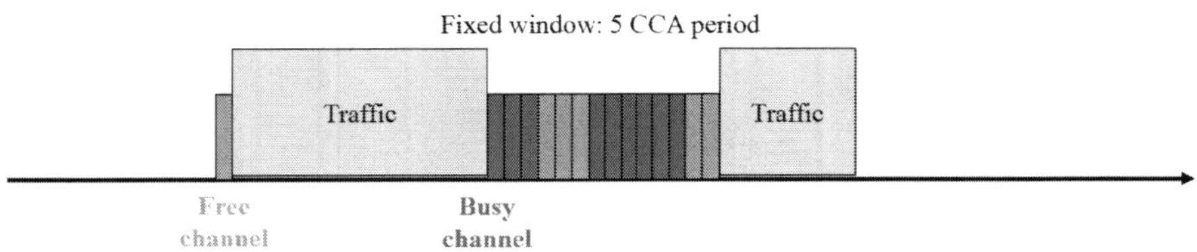

Figure 11.12. *LBT mechanism –LBE option*

The LBE option is relatively similar to the backoff mechanism of Wi-Fi access. Unlike Wi-Fi access, which adopts an exponential backoff mechanism, the LBE option opts for a backoff mechanism with a fixed window.

11.4. PDCP

The Packet Data Convergence Protocol (PDCP) is used for RRC (Radio Resource Control) messages, relating to dedicated control data, and IP (Internet Protocol) packet related to the traffic.

The PDCP performs the following functions:

– compression of traffic data headers using the ROHC (Robust Header Compression) mechanism;

– security of traffic data (confidentiality) and of RRC messages (integrity and confidentiality);

– delivery in sequence of RRC messages and IP packets;

– recovery of PDCP frames lost during the handover.

The PDCP defines headers to encapsulate the RRC signaling data, the traffic data and the control messages associated with the traffic data.

The structure of PDCP frames is described in Figure 11.13 for frames containing traffic data and RRC signaling data.

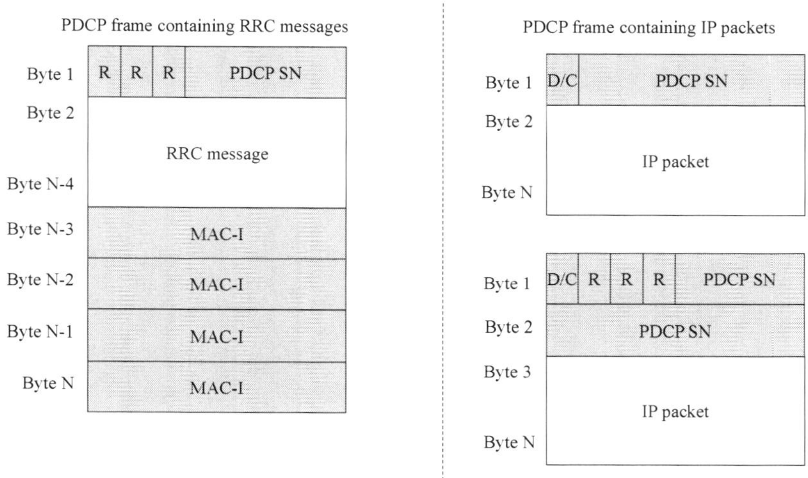

Figure 11.13. *PDCP frame structure containing IP packets or RRC messages*

PDCP SN: this field is coded on five bits for the RRC signaling data and on seven or 12 bits for the traffic data. It indicates the sequence number of the PDCP frame. This sequence number makes it possible to recover the PDCP blocks lost during the handover. For LWA, this field allows the data received from LTE and Wi-Fi access to be put in order.

MAC-I: this field is coded on four bytes. It contains the seal for controlling the integrity of the PDCP frame containing RRC signaling data.

D/C (Data/Control): this bit indicates whether the frame contains traffic data (bit to ONE) or control messages specific to the PDCP (bit to ZERO).

The structure of the PDCP frames is described in Figure 11.14 for frames containing control messages of the PDCP layer.

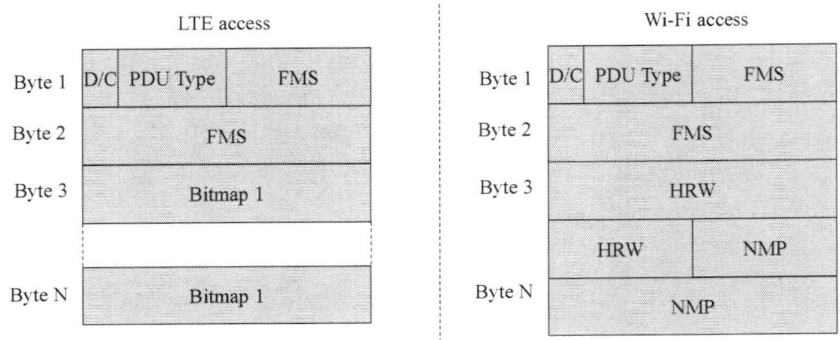

Figure 11.14. *PDCP frame structure containing Status Report messages*

PDU (Packet Data Unit) Type: this field is coded on three bits. It indicates the type of control message associated with the traffic data:

– Status Report message (value 000) is used differently for LTE access and Wi-Fi access. For LTE access, during the recovery procedure following a cell change, it indicates a list of PDCP frame numbers not received during the handover. For Wi-Fi access, it provides error statistics;

– ROHC Feedback message (value 001) is related to the compression mechanism of the headers.

FMS (First Missing): this field is coded on 12, 15 or 18 bits. It contains the first sequence number of the missing PDCP frames.

Bitmap: this field is a collection of bits indicating whether the PDCP frame was received correctly (bit to ONE) or not (bit to ZERO). The most significant bit of the first byte represents the sequence number following the value of the FMS field.

HRW (Highest Received PDCP SN on WLAN): this field is coded on 12, 15 or 18 bits. It contains the highest value of the PDCP SN parameter of the PDCP frame received on the Wi-Fi interface.

NMP (Number of Missing PDCP SDUs): this field is coded on 12, 15 or 18 bits. It contains the number of missing frames from and including the value corresponding to the FMS parameter up to the value of the HRW parameter.

12

MPTCP Aggregation

12.1. Functional architecture

In the case of LWA/LWIP aggregation solutions, the LTE (Long-Term Evolution) and Wi-Fi access aggregation is done at the eNB entity level. In MPTCP-based (Multi-Path Transmission Control Protocol) aggregation solutions, aggregation occurs at the TCP layer.

The target of MPTCP aggregation is to transmit data using multiple paths without causing modifications to existing infrastructures (the 4G mobile network, Wi-Fi access).

The MPTCP connection is performed by the MPTCP client hosted in the mobile and the MPTCP server hosted in an MPTCP proxy. The MPTCP connection is built on TCP connections, each corresponding to an access, LTE or Wi-Fi (Figure 12.1).

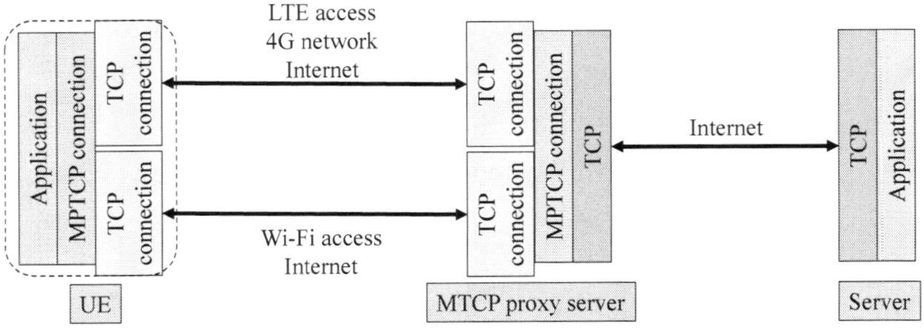

Figure 12.1. *Architecture for MPTCP aggregation*

12.2. TCP

The TCP (Transmission Control Protocol) is a reliable protocol for the following reasons:

– data transmission is conditional upon the establishment of a connection between the source and destination;

– the receiver delivers ordered, error-free data to the application layer;

– the receiver implements a source flow control mechanism based on its receive buffer occupancy;

– the source rate is regulated based on the congestion state of the network.

The TCP is byte-stream oriented:

– at the source side, the application writes bytes into the transmission buffer. The TCP transmits segments to the recipient;

– at the destination side, the application reads bytes into the reception buffer. The application is in charge of delimiting messages.

12.2.1. *TCP header*

The TCP header is described in Figure 12.2 and contains the following fields:

Source Port and Destination Port: these fields identify the port numbers of the source and destination applications.

Sequence Number: this field specifies the current number assigned to the first octet of the data encapsulated by the TCP header. Each byte of the message encapsulated by the TCP header consumes a number.

Acknowledgment Number: this field contains the next sequence number expected by the transmitter of this acknowledgment. It implicitly acknowledges the previous numbers.

Data Offset: this field indicates the number of 32-bit words in the TCP header. Taking into account the presence of the Option field, this field indicates the starting point of the data encapsulated by the TCP header.

Reserved: this field is reserved for later use.

```
 0                   1                   2                   3
 0 1 2 3 4 5 6 7 8 9 0 1 2 3 4 5 6 7 8 9 0 1 2 3 4 5 6 7 8 9 0 1
+-+-+-+-+-+-+-+-+-+-+-+-+-+-+-+-+-+-+-+-+-+-+-+-+-+-+-+-+-+-+-+-+
|          Source Port          |       Destination Port        |
+-+-+-+-+-+-+-+-+-+-+-+-+-+-+-+-+-+-+-+-+-+-+-+-+-+-+-+-+-+-+-+-+
|                        Sequence Number                        |
+-+-+-+-+-+-+-+-+-+-+-+-+-+-+-+-+-+-+-+-+-+-+-+-+-+-+-+-+-+-+-+-+
|                    Acknowledgment Number                      |
+-+-+-+-+-+-+-+-+-+-+-+-+-+-+-+-+-+-+-+-+-+-+-+-+-+-+-+-+-+-+-+-+
|Data   |           |U|A|P|R|S|F|                               |
|Offset | Reserved  |R|C|S|S|Y|I|            Window             |
|       |           |G|K|H|T|N|N|                               |
+-+-+-+-+-+-+-+-+-+-+-+-+-+-+-+-+-+-+-+-+-+-+-+-+-+-+-+-+-+-+-+-+
|           Checksum            |         Urgent Pointer        |
+-+-+-+-+-+-+-+-+-+-+-+-+-+-+-+-+-+-+-+-+-+-+-+-+-+-+-+-+-+-+-+-+
```

Figure 12.2. *Format of TCP header*

Flags:

– URG bit is set to ONE if the Urgent Pointer field is being used;

– ACK bit is set to ONE to indicate the validity of the Acknowledgment Number. When the ACK bit is set to ZERO, the segment does not contain an acknowledgment and the Acknowledgment Number field is ignored;

– PSH bit is set to ONE to notify the recipient to deliver the data to the respective application without buffering;

– RST bit is set to ONE to allow a connection to be reset; it is also used to refuse an attempt to open a connection;

– SYN bit is set to ONE to establish a connection;

– FIN bit is set to ONE to a connection.

Window: this field specifies the window size. The amount of data that the receiver of the segment is able to receive is mentioned in each segment. This field enables flow control to be performed. The receiver may respond with a zero-valued Window field. This indicates that its reception buffer is full.

Checksum: this field is used to detect whether bit errors occurred during transfer. This check encompasses the header, the data encapsulated by the TCP header and a pseudo-header. The pseudo-header includes the source and destination IP addresses, the Protocol field from the IP header and the length of the IP packet.

Urgent Pointer: this field points to the first byte where urgent data can be found. It allows the sender to transmit information to the receiver without interrupting the message transmission in progress.

Options: this field may be used for various functions:

– maximum segment size (MSS): this information is only provided during the connection opening procedure. With an MTU of 1,500 bytes over Ethernet, the MSS is 1,460 bytes in IPv4 and 1,440 bytes in IPv6;

– sizing of the reception buffer (Window Scale): the sender and receiver may negotiate a scale factor for the window size. This option makes it possible to circumvent the limitation in window size due to the 16-bit length of the Window field;

– selective acknowledgment (SACK) instead of a bulk retransmission: the receiver may request retransmission of one or more specific segments. After obtaining them, it can then acknowledge all the buffered data. This option provides two pointers to indicate each received data block.

12.2.2. *Opening and closing a connection*

Opening a connection involves the following three steps:

– the client sends a segment having a SYN bit set to ONE and an ACK bit set to ZERO. The client indicates the sequence number X;

– the server responds with the SYN bit set to ONE and the ACK bit set to ONE. The server indicates the sequence number Y. The server validates the received segment by indicating the next expected byte X + 1;

– the client responds with the SYN bit set to ONE and the ACK bit set to ONE. The client indicates the sequence number X and validates the received segment by indicating the next expected byte Y + 1.

An endpoint can close the connection using either the FIN bit or the RST bit. The FIN bit allows the connection to be closed when the data to be transferred have actually been received by the recipient. The RST bit is used to close the session abruptly.

12.2.3. *Data transfer*

The TCP has the Sequence Number, Acknowledgment Number and Checksum fields to perform error-free data transfer. At the time a segment is transmitted, a retransmission timeout (RTO) is initiated.

If the segment is acknowledged before the expiration of the timer, then the latter is stopped. On the other hand, if the timer expires before the acknowledgment arrives, then the segment is retransmitted and the timer is activated again.

When a segment is retransmitted, the retransmission timer is not updated, its value being doubled instead.

In a standard configuration, retransmission resumes from the segment for which the timer expired, since the sender has no information as to whether or not the next segments were actually received, unless the SACK option is used.

The TCP continuously adjusts the value of this retransmission timer. It manages an RTT (Round Trip Time) variable, which is the current estimate of the time elapsed between the transmission of a segment and the reception of the acknowledgment.

12.2.4. *Slow Start and Congestion Avoidance mechanisms*

The source takes advantage of the Slow Start and Congestion Avoidance mechanisms to control the amount of transmitted data. These mechanisms are implemented using two variables:

– the congestion window (cwnd) indicates the amount of bytes the source can transmit before it receives an acknowledgment via the Acknowledgment Number field;

– the receive window (rwnd) indicates the amount of bytes the receiver is ready to receive, a value indicated via the Window field.

On establishing the connection, the sender initializes the congestion window (cwnd) with a value equal to two segments of maximum length (MSS) (Figure 12.3). If the segments are acknowledged before the expiration of the timer, the size of the congestion window is doubled. Each successful transmission thus enables the size of the congestion window to be doubled.

The Congestion Avoidance mechanism introduces a new variable: the congestion avoidance threshold (ssthresh) (Figure 12.3). When the congestion avoidance threshold is reached, the congestion window grows linearly by an increment corresponding to the maximum size of one segment.

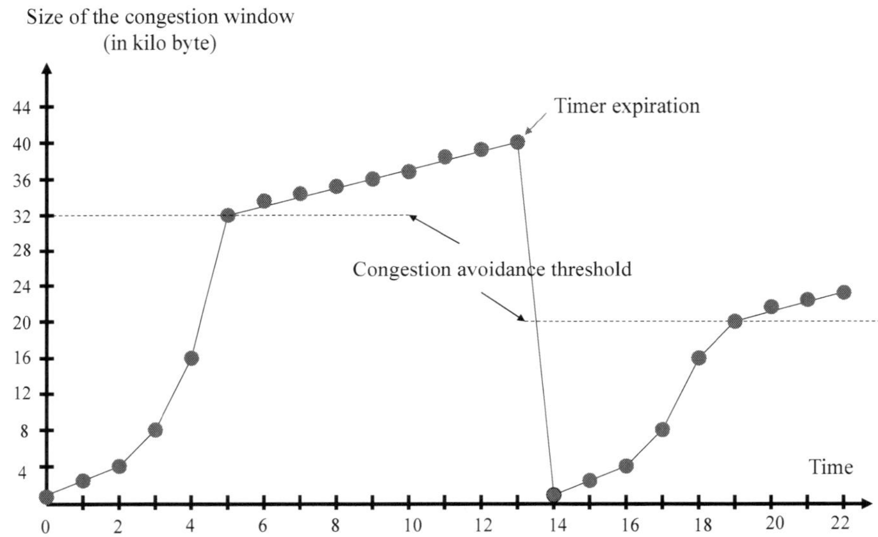

Figure 12.3. *Slow Start and Congestion Avoidance mechanisms*

In the beginning, the value of the congestion avoidance threshold is equal to half that of the receive window (rwnd). If the timer expires before receiving the acknowledgment for the transmitted data, then the congestion avoidance threshold is then set to half the current congestion window and the Slow Start mechanism restarts.

12.2.5. *Fast Retransmit and Fast Recovery mechanisms*

The Fast Retransmit mechanism enables the sender to retransmit a segment without waiting for the timer to expire, thus avoiding the start-up of the Slow Start mechanism (Figure 12.4).

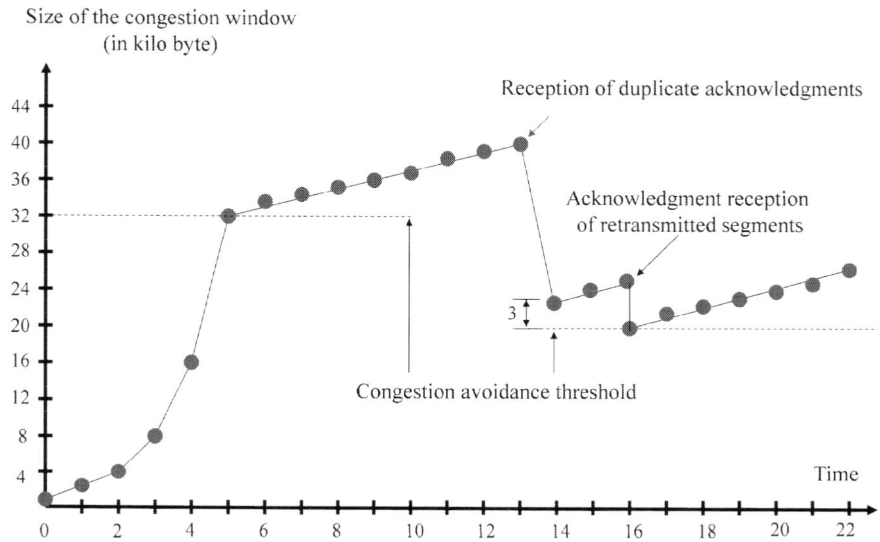

Figure 12.4. *Fast Retransmit and Fast Recovery mechanisms*

If a segment is missing at the receiver, then the latter will emit segments with duplicate acknowledgments, having the same value in the Acknowledgment Number field.

At the other end, the reception of duplicate acknowledgments is used to determine the missing segment. It should be noted that the reception of duplicate acknowledgments may also be caused by network-induced desequencing. When three duplicate acknowledgments are received, the sender resends the missing segment.

The Fast Recovery mechanism operates as follows:

– when three duplicate acknowledgments are received, the congestion avoidance threshold (ssthresh) is set to half the current congestion window;

– the congestion window (cwnd) takes on a value greater than the congestion avoidance threshold by three MSS values;

– for every additional duplicate acknowledgment received, the congestion window is incremented by one MSS value;

– at the first acknowledgment corresponding to the retransmitted segments, the congestion window is positioned at the congestion avoidance threshold.

12.2.6. *ECN mechanism*

WRED (Weighed Random Early Discard) queue management is a method of giving the recipient an indication of congestion. It is based on the anticipated and random destruction of packets.

The ECN (Explicit Congestion Notification) mechanism constitutes another method to provide such a congestion indication without destroying packets. It is based on the use of fields in the header of the IP and TCP to warn terminal stations about the beginning of congestion.

The ECN mechanism uses two bits in the IP header (Figure 12.5). These two bits complement the six bits in the DSCP (DiffServ Code Point) field.

```
  0                   1                   2                   3
  0 1 2 3 4 5 6 7 8 9 0 1 2 3 4 5 6 7 8 9 0 1 2 3 4 5 6 7 8 9 0 1
 +-+-+-+-+-+-+-+-+-+-+-+-+-+-+-+-+-+-+-+-+-+-+-+-+-+-+-+-+-+-+-+-+
 |Version|  IHL  |  DiffServ |ECN|          Total Length         |
 +-+-+-+-+-+-+-+-+-+-+-+-+-+-+-+-+-+-+-+-+-+-+-+-+-+-+-+-+-+-+-+-+
```

Figure 12.5. *ECN field in IP header*

When these two bits equal 10 or 01, they indicate that the source and the destination are able to treat the ECN mechanism. These two bits are called ECT (ECN-Capable Transport, ECT(0) and ECT(1) respectively). Both endpoints are made aware of each other's ability to treat the ECN mechanism during the TCP connection procedure (Table 12.1).

ECN field		Designation
0	0	Non-ECT
0	1	ECT(1)
1	0	ECT(0)
1	1	CE

Table 12.1. *ECN field in IP header*

When these two bits equal 00, they indicate that the source and the destination are not able to treat the ECN mechanism (Table 12.1).

When the router detects the beginning of congestion through the WRED queue management mechanism, the two bits assume the value 11 (Table 12.1). These two bits are called CE (Congestion Experienced).

The ECN field only assumes the CE value if this field previously had the value ECT(0) or ECT(1) or if this field was already flagged as CE. Should the ECN field equal 00, the detection of incipient congestion would result in packet destruction. The CE value is used by the router to notify the recipient of congestion onset.

The ECN mechanism further comprises two bits in the TCP header (Figure 12.6). These two bits complement the six flag bits.

```
 0                   1                   2                   3
 0 1 2 3 4 5 6 7 8 9 0 1 2 3 4 5 6 7 8 9 0 1 2 3 4 5 6 7 8 9 0 1
+-+-+-+-+-+-+-+-+-+-+-+-+-+-+-+-+-+-+-+-+-+-+-+-+-+-+-+-+-+-+-+-+
|Data   |       |C|E|U|A|P|R|S|F|                               |
|Offset | Res   |W|C|R|C|S|S|Y|I|            Window             |
|       |       |R|E|G|K|H|T|N|N|                               |
+-+-+-+-+-+-+-+-+-+-+-+-+-+-+-+-+-+-+-+-+-+-+-+-+-+-+-+-+-+-+-+-+
```

Figure 12.6. *ECN field in TCP header*

The ECE bit (ECN-Echo) is exploited by the recipient to warn the source that a beginning of congestion has been detected in the network. This bit is set when the recipient receives a packet whose ECN field in the IP header has the CE value.

The CWR (Congestion Window Reduced) bit is used by the source to notify the recipient that it has indeed received the ECE flag. When the recipient receives the CWR flag, it stops transmitting the ECE flag.

While the TCP connection is being opened, the client sends a TCP header with the SYN (connection opening), ECE and CWR flags set to one. The server responds with a TCP header, wherein the SYN (connection opening) and ECE flags are set to one. During that stage, the ECE and CWR bits are used to indicate the ability of terminals to treat the ECN mechanism, rather than the occurrence of congestion in the network.

12.3. MPTCP

The MPTCP connection consists of a combination of several TCP connections, with each TCP connection transmitted on a path to appear as a normal connection for the different devices crossed:

– each TCP connection must start with the establishment procedure;

– each TCP connection must independently manage its sequence number;

– each TCP connection must end with FIN or RST flag.

The MPTCP uses the Kind (= 30) option of the TCP header, with each MPTCP message being identified by the Subtype field (Figure 12.7).

```
                     1                   2                   3
 0 1 2 3 4 5 6 7 8 9 0 1 2 3 4 5 6 7 8 9 0 1 2 3 4 5 6 7 8 9 0 1
+---------------+---------------+-------+-----------------------+
|     Kind      |    Length     |Subtype|                       |
+---------------+---------------+-------+                       |
|                     Subtype-specific data                     |
|                       (variable length)                       |
----------------------------------------------------------------+
```

Figure 12.7. *Format of MPTCP option*

Table 12.2 summarizes the different MPTCP options corresponding to the Subtype field.

Value (in hexadecimal)	Designation
00	MP_CAPABLE
01	MP_JOIN
02	DSS
03	ADD_ADDR
04	REMOVE_ADDR
05	MP_PRIO
06	MP_FAIL
07	MP_FASTCLOSE

Table 12.2. *MPTCP options*

12.3.1. *Establishment of MPTCP connection*

The establishment of MPTCP connection begins with an exchange SYN, SYN/ACK, ACK of the first TCP connection. Each TCP header contains the MP_CAPABLE option. This option declares that the sender is capable of rolling MPTCP over the first TCP connection (Figure 12.8).

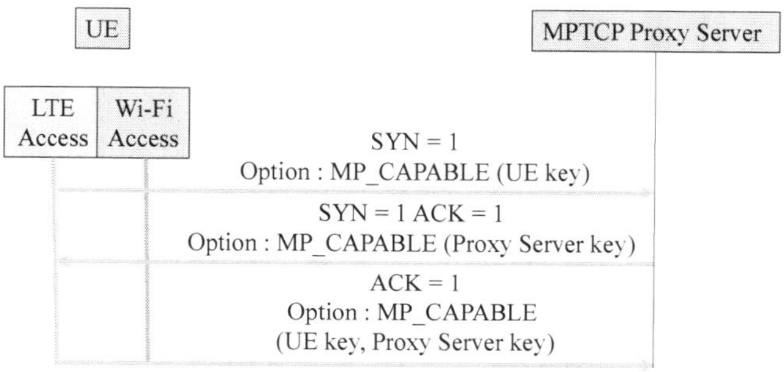

Figure 12.8. *Establishment of an MPTCP connection*

This option is used to declare the 64-bit key that each entity (the mobile, the Proxy Server MPTCP) has generated for this MPTCP connection. This key will later authenticate the addition of new TCP connections to this MPTCP connection.

This is the only time that the key will be sent clearly on the first TCP connection, with the exception of the MP_FASTCLOSE option. All future TCP connections will identify the connection using a 32-bit token. This token is a cryptographic hash of this key.

12.3.2. *Adding a TCP connection*

When the MPTCP connection is established with the MP_CAPABLE exchange on the LTE access, a new TCP connection can be added for the Wi-Fi access.

The new TCP connection starts with a normal exchange SYN, SYN/ACK, ACK. The MP_JOIN option is used to identify the new TCP connection (Figure 12.9).

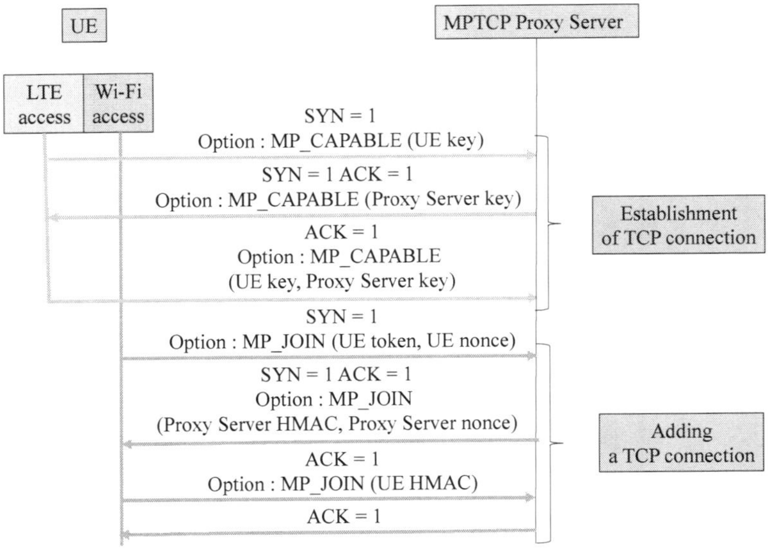

Figure 12.9. *Adding a TCP connection*

The mobile transmits the SYN flag in the first TCP header whose MP_JOIN option contains a token and a random number (nonce).

The token is a cryptographic hash generated from the key of the MPTCP Proxy Server. It is used by the MPTCP Proxy Server to identify the MPTCP connection.

The random number (nonce) is used to calculate the HMAC authentication of the following headers:

The MPTCP Proxy Server responds with the SYN/ACK flags in the TCP header whose MP_JOIN option contains HMAC (Proxy Server) authentication and a random number (nonce):

HMAC (Proxy Server) = HMAC (Key, Message)

Key = Proxy Server key+ mobile key

Message = nonce (Proxy Server) + nonce (mobile)

The mobile performs the third exchange with the ACK flag in the TCP header whose MP_JOIN option contains HMAC (mobile) authentication:

HMAC (mobile) = HMAC (Key, Message)

Key = mobile key + Proxy Server key

Message = nonce (mobile) + nonce (Proxy Server)

The Proxy Server MPTCP terminates the procedure with the ACK flag in the TCP header to acknowledge the TCP header received during the third exchange.

Each MP_JOIN option contains an address identifier. The address identifier has only a meaning in a single connection, where it identifies the source IP address of this packet, even if this IP address has been modified by an intermediate device (e.g. by the NAT function).

Hosts can specify in the MP_JOIN option when configuring a TCP connection whether they want the TCP connection to be used as a normal path or as a backup path. The host may request a change in the priority of the connection via the MP_PRIO option.

12.3.3. *Data transfer*

The DSS (Data Sequence Signal) option provides control information to enable the reassembly of streams from different TCP connections and its reliable and orderly delivery of data for the application at the destination level.

To deliver the data in sequence, the DSS option contains the mapping between the data sequence number (DSN) of the MPTCP connection and the sub-flow sequence number (SSN) of the TCP connection (Figure 12.10).

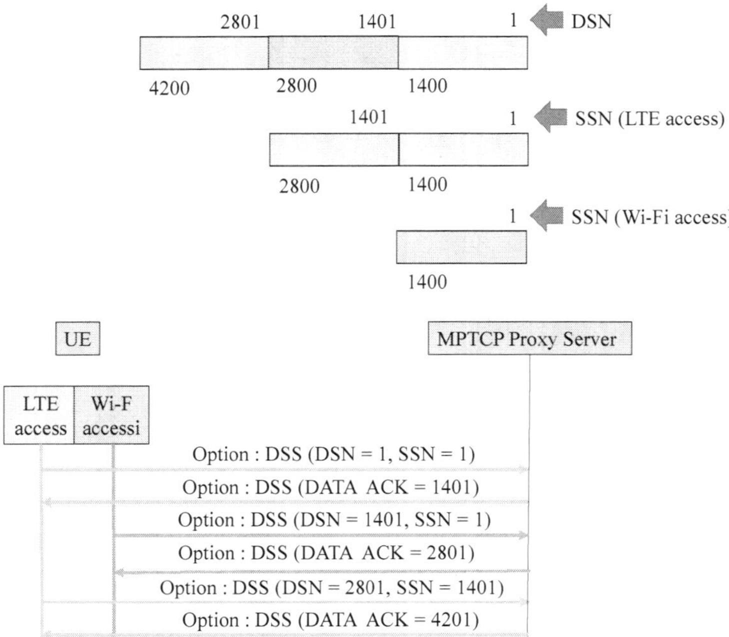

Figure 12.10. *Data transfer*

The DSS option uses the Data ACK field to acknowledge data from the MPTCP connection. This acknowledgment indicates to the sender that the data corresponding to the DSN number has been received and specifies the next DSN to be transmitted.

For a normal TCP connection, the recipient of a segment notifies the source of the available size for the receive memory in the Window field.

The MPTCP connection uses a single receive window for all TCP connections, which allows each TCP connection to transmit data if the size of the receive memory allows.

The value of the receive window is indicated in the Window field of the TCP connection, whose associated DSS option contains in the DATA_ACK field the number of the next expected byte relative to the DSN sequence (Figure 12.10). The source can transmit on all TCP connections segments whose DSN number is between DATA_ACK and DATA_ACK + Window.

For a normal TCP connection, congestion control introduces another limitation provided by the congestion window (cwnd).

Executing the existing algorithms independently for each TCP connection would give very high flow rate for MPTCP connection if these TCP connections cross a single bottleneck.

In addition, it is desirable for a source with multiple available accesses to use the least congested path.

The congestion control algorithm applies only to the increase phase of the reception window. For this, it is necessary to couple the congestion control algorithms operating on the different TCP connections and dynamically control the overall aggressiveness of the MPTCP connection.

The Slow Start, Fast Retransmit and Fast Recovery mechanisms are the same as for the standard TCP.

When a retransmission over a TCP connection fails, the sender can send the data from the MPTCP connection with the same DSN sequence number to another TCP connection after a timeout.

During a retransmission, the host must, however, try to retransmit the original data over the original TCP connection, in order to preserve the integrity of the SSN sequence numbers.

12.3.4. *Closing an MPTCP connection*

When an application requests the closing of a session, it indicates that it no longer has data to send. For a TCP connection, this causes the FIN flag to be issued. For an MPTCP connection, an equivalent mechanism is required, based on the DATA_FIN indication contained in the DSS option (Figure 12.11).

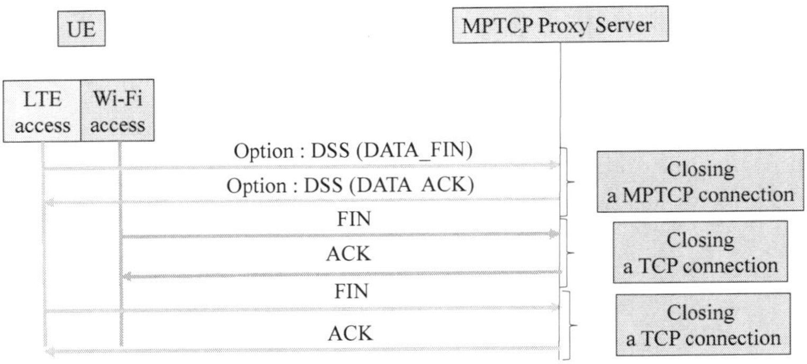

Figure 12.11. *Closing a MPTCP connection*

The closure of the MPTCP connection is decoupled from the closures of the TCP connections. When DATA_FIN has been acknowledged, all remaining TCP connections must be closed with exchanges of the FIN flag. The DATA_FIN indication is acknowledged when all data has been transferred to the different TCP connections.

The TCP connection has the means to close a connection with the RST flag abruptly. The option MP_FASTCLOSE allows the same function to be performed for the MPTCP connection (Figure 12.12).

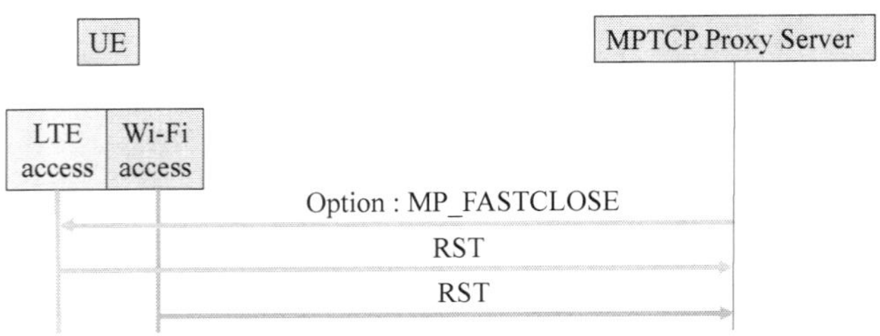

Figure 12.12. *Abrupt closure of MPTCP connection*

Host A (e.g. the proxy server) sends, over a TCP connection, a TCP header containing the MP_FASTCLOSE option, containing the key of host B (for example the mobile) declared during the establishment of the MPTCP connection. For all TCP connections, host A sends the RST flag.

On receipt of the MP_FASTCLOSE option containing the valid key, host B responds on the same TCP connection with the RST flag and closes the other TCP connections.

12.3.5. *Adding and removing an address*

An MPTCP connection is initially configured between the A1 pair (address/port for LTE access) of host A (the mobile) and the B1 pair (address/port) of host B (the proxy server). Host A can start an additional TCP connection from its A2 pair (Wi-Fi access) by sending the SYN flag and the MP_JOIN option.

At the end of the first TCP connection, host A uses the ADD_ADDR option, informing the recipient of the A2 pair (address/port).

Each IP address is associated with a unique address identifier (address ID) that can be used for address removal or for the MP_JOIN option.

Due to the proliferation of NAT devices, it is likely that a host might attempt to publish private IP addresses. The MP_JOIN procedure for creating a new TCP connection provides mechanisms to minimize security risks.

Similarly, host B can use the ADD_ADDR option to inform host A of the availability of the B2 pair (address/port).

From the point of view of host A (the mobile), the following four TCP connections can thus be opened:

address/port A1 and address/port B1, on LTE access;

address/port A1 and address/port B2, on LTE access;

address/port A2 and address/port B1, on Wi-Fi access;

address/port A2 and address/port B2, on Wi-Fi access.

If during an MPTCP connection, a previously advertised address becomes invalid, then the affected host must advertise it in the REMOVE_ADDR option so that the other end can delete the TCP connections bound to that address.

12.3.6. *Return to the TCP connection*

At the beginning of an MPTCP connection, it is important to ensure that the path is fully compatible with the MPTCP options. If any of the SYN flags do not have the MPTCP options, the session must continue automatically on a regular TCP connection.

This scenario occurs when the host does not process MPTCP options or when intermediate devices prohibit the use of options.

In the case where the data is not contiguous, which could happen when a single TCP connection retransmits data from another recently closed TCP connection, the recipient must close the TCP connection with the RST flag and the MP_FAIL option. The recipient must reject all data that follow the specified DSN sequence number. The source may try to restart the TCP connection.

Bibliography

General Documentation

[3GPP TS 23.402] Architecture enhancements for non-3GPP accesses – 2016.

[3GPP TS 24.302] Access to the 3GPP Evolved Packet Core (EPC) via non-3GPP access networks – 2016.

[3GPP TS 24.234] WLAN User Equipment (WLAN UE) to network protocols – 2015.

[3GPP TS 33.234] Wireless Local Area Network (WLAN) interworking security – 2016.

[3GPP TS 33.402] 3GPP System Architecture Evolution (SAE); Security aspects of non-3GPP accesses – 2014.

Integration of Cellular and Wi-Fi Networks – 4G Americas – 2013.

Wi-Fi Roaming Guidelines – GSMA – IR.61 – 2015.

WLAN Traffic Offload in LTE – A. Schumacher, J. Schlienz – Rohde & Schwarz – 2012.

Analysis of LTE / Wi-Fi Aggregation Solutions – Netmanias – 2016.

Network Architecture for LTE and Wi-Fi Interworking – Netmanias – C. Yoo – 2012.

Chapter 1: Architecture Based on Wi-Fi Interface

[3GPP TS 23.203] Policy and charging control architecture – 2016.

[3GPP TS 29.273] Evolved Packet System (EPS); 3GPP EPS AAA interfaces – 2016.

Chapter 2: MAC Layer

[IEEE 802.11] Wireless LAN Medium Access Control (MAC) and Physical Layer (PHY) Specifications – 2012.

802.11 Wireless Network: The definitive guide – M. Gast – O'Reilly – 2005.

[NIST 800-97] Establishing Wireless Robust Security Networks: A Guide to IEEE 802.11i – S. Frankel, B. Eydt, L. Owens, K. Scarfone – 2007.

Chapter 3: 802.11a/g Interfaces

[IEEE 802.11] Wireless LAN Medium Access Control (MAC) and Physical Layer (PHY) Specifications – Chapters 18 / 19 – 2012.

802.11 Wireless Network: The definitive guide – M. Gast – O'Reilly – Chapters 13 / 14 – 2005.

Chapter 4: 802.11n Interface

[IEEE 802.11] Wireless LAN Medium Access Control (MAC) and Physical Layer (PHY) Specifications – Chapter 20 – 2012.

802.11n A survival guide – M. Gast – O'Reilly – 2012.

WLAN 802.11n: From SISO to MIMO – D. Liebl – Rohde & Schwarz – 2012.

Chapter 5: 802.11ac Interface

[IEEE 802.11] Wireless LAN Medium Access Control (MAC) and Physical Layer (PHY) Specifications - Amendment 4: Enhancements for Very High Throughput for Operation in Bands below 6 GHz – 2013.

802.11ac A survival guide – M. Gast – O'Reilly – 2013.

802.11ac Technology Introduction – L. Ward – Rohde & Schwarz – 2012.

802.11ac: The Fifth Generation of Wi-Fi – Cisco – 2014.

Chapter 6: Mutual Authentication

[IEEE 802.1X] Port-based Network Access Control – 2010.

[RFC 3748] Extensible Authentication Protocol (EAP) – B. Aboba, L. Blunk, J. Vollbrecht, J. Carlson, H. Levkowetz – 2004.

[RFC 4187] Extensible Authentication Protocol Method for 3rd Generation; Authentication and Key Agreement (EAP-AKA) – J. Arkko, H. Haverinen – 2006.

[RFC 5448] Improved Extensible Authentication Protocol Method for 3rd Generation; Authentication and Key Agreement (EAP-AKA') – J. Arkko, V. Lehtovirta, P. Eronen – 2009.

[RFC 2865] Remote Authentication Dial In User Service (RADIUS) – C. Rigney, S. Willens, A. Rubens, W. Simpson – 2000.

[RFC 3579] RADIUS (Remote Authentication Dial In User Service) Support For Extensible Authentication Protocol (EAP) – B. Aboba, P. Calhoun – 2003.

Chapter 7: SWu Tunnel Establishment

[RFC 4301] Security Architecture for the Internet Protocol – S. Kent, K. Seo – 2005.

[RFC 4303] IP Encapsulating Security Payload (ESP) – S. Kent – 2005.

[RFC 5996] Internet Key Exchange Protocol Version 2 (IKEv2) – C. Kaufman, P. Hoffman, Y. Nir, P. Eronen – 2010.

[RFC 4555] IKEv2 Mobility and Multihoming Protocol (MOBIKE) – P. Eronen – 2006.

[NIST 800-77] Guide to IPsec VPNs – S. Frankel, K. Kent, R. Lewkowski, A.D. Orebaugh, R.W. Ritchey, S.R. Sharma – 2005.

Chapter 8: S2a / S2b Tunnel Establishment

[RFC 5213] Proxy Mobile IPv6 – S. Gundavelli, K. Leung, V. Devarapalli, K. Chowdhury, B. Patil – 2008.

[3GPP TS 29.275] Proxy Mobile IPv6 (PMIPv6) based Mobility and Tunnelling protocols – 2016.

[3GPP TS 29.274] Evolved General Packet Radio Service (GPRS) Tunnelling Protocol for Control plane (GTPv2-C) – 2016.

[RFC 5944] IP Mobility Support for IPv4, Revised – C. Perkins – 2010.

[3GPP TS 29.279] Mobile IPv4 (MIPv4) based Mobility protocols – 2011.

Chapter 9: S2c Tunnel Establishment

[RFC 6275] Mobility Support in IPv6 – C. Perkins, D. Johnson, J. Arkko – 2011.

[RFC 5555] Mobile IPv6 Support for Dual Stack Hosts and Routers – H. Soliman – 2009.

Chapter 10: Network Discovery and Selection

[3GPP TS 24.312] Access Network Discovery and Selection Function (ANDSF); Management Object (MO) – 2016.

Hotspot 2.0 (Release 2) Technical Specification – Wi-Fi Alliance – 2016.

Chapter 11: Carrier Aggregation

[3GPP TS 36.300] Evolved Universal Terrestrial Radio Access (E-UTRA) and Evolved Universal Terrestrial Radio Access Network (E-UTRAN); Overall description – 2017.

[3GPP TS 36.323] Packet Data Convergence Protocol (PDCP) specification – 2017.

LTE Aggregation & Unlicensed Spectrum – 5G Americas – 2015.

Chapter 12: MPTCP Aggregation

[RFC 6182] Architectural Guidelines for Multipath TCP Development – A. Ford, C. Raiciu, M. Handley, S. Barre, J. Iyengar – 2011.

[RFC 6824] TCP Extensions for Multipath Operation with Multiple Addresses – A. Ford, C. Raiciu, M. Handley, O. Bonaventure – 2013.

Index

D, E, F

G, H, I

K, L, M

N, O, P

Q, R, S

T, U, V, W

Other titles from

ISTE

in

Networks and Telecommunications

2018

ANDIA Gianfranco, DURO Yvan, TEDJINI Smail
Non-linearities in Passive RFID Systems: Third Harmonic Concept and Applications

2017

BENSLAMA Malek, BENSLAMA Achour, ARIS Skander
Quantum Communications in New Telecommunications Systems

HILT Benoit, BERBINEAU Marion, VINEL Alexey, PIROVANO Alain
Networking Simulation for Intelligent Transportation Systems: High Mobile Wireless Nodes

LESAS Anne-Marie, MIRANDA Serge
The Art and Science of NFC Programming
(Intellectual Technologies Set – Volume 3)

2016

AL AGHA Khaldoun, PUJOLLE Guy, ALI-YAHIYA Tara
Mobile and Wireless Networks
(Advanced Network Set – Volume 2)

BATTU Daniel
Communication Networks Economy

BENSLAMA Malek, BATATIA Hadj, MESSAI Abderraouf
Transitions from Digital Communications to Quantum Communications: Concepts and Prospects

CHIASSERINI Carla Fabiana, GRIBAUDO Marco, MANINI Daniele
Analytical Modeling of Wireless Communication Systems
(Stochastic Models in Computer Science and Telecommunication Networks Set – Volume 1)

EL FALLAH SEGHROUCHNI Amal, ISHIKAWA Fuyuki, HÉRAULT Laurent, TOKUDA Hideyuki
Enablers for Smart Cities

PEREZ André
VoLTE and ViLTE

2015

BENSLAMA Malek, KIAMOUCHE Wassila, BATATIA Hadj
Connections Management Strategies in Satellite Cellular Networks

BENSLAMA Malek, BATATIA Hadj, BOUCENNA Mohamed Lamine
Ad Hoc Networks Telecommunications and Game Theory

BERTHOU Pascal, BAUDOIN Cédric, GAYRAUD Thierry, GINESTE Matthieu
Satellite and Terrestrial Hybrid Networks

CUADRA-SANCHEZ Antonio, ARACIL Javier
Traffic Anomaly Detection

LE RUYET Didier, PISCHELLA Mylène
Digital Communications 1: Source and Channel Coding

PEREZ André
LTE and LTE Advanced: 4G Network Radio Interface

PISCHELLA Mylène, LE RUYET Didier
Digital Communications 2: Digital Modulations

PUJOLLE Guy
Software Networks
(Advanced Network Set – Volume 1)

2014

ANJUM Bushra, PERROS Harry
Bandwidth Allocation for Video under Quality of Service Constraints

BATTU Daniel
New Telecom Networks: Enterprises and Security

BEN MAHMOUD Mohamed Slim, GUERBER Christophe, LARRIEU Nicolas, PIROVANO Alain, RADZIK José
Aeronautical Air−Ground Data Link Communications

BITAM Salim, MELLOUK Abdelhamid
Bio-inspired Routing Protocols for Vehicular Ad-Hoc Networks

CAMPISTA Miguel Elias Mitre, RUBINSTEIN Marcelo Gonçalves
Advanced Routing Protocols for Wireless Networks

CHETTO Maryline
Real-time Systems Scheduling 1: Fundamentals
Real-time Systems Scheduling 2: Focuses

EXPOSITO Ernesto, DIOP Codé
Smart SOA Platforms in Cloud Computing Architectures

MELLOUK Abdelhamid, CUADRA-SANCHEZ Antonio
Quality of Experience Engineering for Customer Added Value Services

OTEAFY Sharief M.A., HASSANEIN Hossam S.
Dynamic Wireless Sensor Networks

PEREZ André
Network Security

PERRET Etienne
Radio Frequency Identification and Sensors: From RFID to Chipless RFID

REMY Jean-Gabriel, LETAMENDIA Charlotte
LTE Standards
LTE Services

TANWIR Savera, PERROS Harry
VBR Video Traffic Models

VAN METER Rodney
Quantum Networking

XIONG Kaiqi
Resource Optimization and Security for Cloud Services

2013

ASSING Dominique, CALÉ Stéphane
Mobile Access Safety: Beyond BYOD

BEN MAHMOUD Mohamed Slim, LARRIEU Nicolas, PIROVANO Alain
Risk Propagation Assessment for Network Security: Application to Airport Communication Network Design

BEYLOT André-Luc, LABIOD Houda
Vehicular Networks: Models and Algorithms

BRITO Gabriel M., VELLOSO Pedro Braconnot, MORAES Igor M.
Information-Centric Networks: A New Paradigm for the Internet

BERTIN Emmanuel, CRESPI Noël
Architecture and Governance for Communication Services

DEUFF Dominique, COSQUER Mathilde
User-Centered Agile Method

DUARTE Otto Carlos, PUJOLLE Guy
Virtual Networks: Pluralistic Approach for the Next Generation of Internet

FOWLER Scott A., MELLOUK Abdelhamid, YAMADA Naomi
LTE-Advanced DRX Mechanism for Power Saving

JOBERT Sébastien *et al.*
Synchronous Ethernet and IEEE 1588 in Telecoms: Next Generation Synchronization Networks

MELLOUK Abdelhamid, HOCEINI Said, TRAN Hai Anh
Quality-of-Experience for Multimedia: Application to Content Delivery Network Architecture

NAIT-SIDI-MOH Ahmed, BAKHOUYA Mohamed, GABER Jaafar, WACK Maxime
Geopositioning and Mobility

PEREZ André
Voice over LTE: EPS and IMS Networks

2012

AL AGHA Khaldoun
Network Coding

BOUCHET Olivier
Wireless Optical Communications

DECREUSEFOND Laurent, MOYAL Pascal
Stochastic Modeling and Analysis of Telecoms Networks

DUFOUR Jean-Yves
Intelligent Video Surveillance Systems

EXPOSITO Ernesto
Advanced Transport Protocols: Designing the Next Generation

JUMIRA Oswald, ZEADALLY Sherali
Energy Efficiency in Wireless Networks

KRIEF Francine
Green Networking

PEREZ André
Mobile Networks Architecture

2011

BONALD Thomas, FEUILLET Mathieu
Network Performance Analysis

CARBOU Romain, DIAZ Michel, EXPOSITO Ernesto, ROMAN Rodrigo
Digital Home Networking

CHABANNE Hervé, URIEN Pascal, SUSINI Jean-Ferdinand
RFID and the Internet of Things

GARDUNO David, DIAZ Michel
Communicating Systems with UML 2: Modeling and Analysis of Network Protocols

LAHEURTE Jean-Marc
Compact Antennas for Wireless Communications and Terminals: Theory and Design

RÉMY Jean-Gabriel, LETAMENDIA Charlotte
Home Area Networks and IPTV

PALICOT Jacques
Radio Engineering: From Software Radio to Cognitive Radio

PEREZ André
IP, Ethernet and MPLS Networks: Resource and Fault Management

TOUTAIN Laurent, MINABURO Ana
Local Networks and the Internet: From Protocols to Interconnection

2010

CHAOUCHI Hakima
The Internet of Things

FRIKHA Mounir
Ad Hoc Networks: Routing, QoS and Optimization

KRIEF Francine
Communicating Embedded Systems / Network Applications

2009

CHAOUCHI Hakima, MAKNAVICIUS Maryline
Wireless and Mobile Network Security

VIVIER Emmanuelle
Radio Resources Management in WiMAX

2008

CHADUC Jean-Marc, POGOREL Gérard
The Radio Spectrum

GAÏTI Dominique
Autonomic Networks

LABIOD Houda
Wireless Ad Hoc and Sensor Networks

LECOY Pierre
Fiber-optic Communications

MELLOUK Abdelhamid
End-to-End Quality of Service Engineering in Next Generation Heterogeneous Networks

PAGANI Pascal *et al.*
Ultra-wideband Radio Propagation Channel

2007

BENSLIMANE Abderrahim
Multimedia Multicast on the Internet

PUJOLLE Guy
Management, Control and Evolution of IP Networks

SANCHEZ Javier, THIOUNE Mamadou
UMTS

VIVIER Guillaume
Reconfigurable Mobile Radio Systems

Printed and bound by CPI Group (UK) Ltd, Croydon, CR0 4YY